Chance Rules

Brian Everitt

Chance Rules

An Informal Guide to Probability, Risk and Statistics

Second Edition

 Springer

Brian Everitt
King's College
School of Medicine
London
UK
brian.everitt@iop.kcl.ac.uk

ISBN: 978-0-387-78129-7   e-ISBN: 978-0-387-77415-2
DOI: 10.1007/978-0-387-77415-2

Library of Congress Control Number: 2008931104

Printed on acid-free paper

springer.com

To the memory of Padmal de Silva,
cricket expert, horserace tipster and much missed friend.

Preface to the Second Edition

Life is a gamble at terrible odds—if it was a bet you wouldn't take it.
Tom Stoppard, *Rosencrantz and Guildenstern are Dead*

Chance continues to rule in the early part of the twenty-first century and probability continues to be the natural language of uncertainty. In this second edition of *Chance Rules* I have added more material on waiting times, some counter-intuitive results when tossing coins or dice, two recent examples of the Prosecutor's Fallacy when dealing with conditional probabilities, extra examples of clinical trials and puzzling probabilities, and a new chapter on predicting the future. As in the first edition, I have tried to keep the mathematical details to the very minimum although a few formidable looking formulae do occasionally appear, particularly in Chapter 10. As with the first edition, it will often be worthwhile to make an effort to follow these more tricky sections, and once again, I wish you luck.

Dulwich, London, 2007   Brian S. Everitt

Acknowledgments to the Second Edition

I am grateful to the Random House Group Ltd for allowing me to reproduce the extract from *The Life and Times of the Thunderbolt kid* by Bill Bryson in Chapter 10. I would like to thank John Kimmel of Springer-Verlag for his constant support and encouragement whilst writing this second edition and for his input of several good ideas that have been incorporated into the book. Thanks are also due to Michael Hayden, tennis player extraordinaire, for help in understanding how stock markets operate when writing Chapter 14.

Preface to the First Edition

Chance governs our lives. From the genes we inherit and the environment into which we are born, to the investments we may make in the stock market or the lottery ticket we buy at the local newsstand, life is a gamble. And whatever our moral persuasion, it is a gamble from which none of us is exempt. In business, education, travel, health and marriage we take chances in the hope of obtaining something better. In an effort to make our children's lives safer, for example, we may choose to have them vaccinated against measles, even though we are aware that there is a small risk that the vaccine will have some dreadful consequence, such as permanent brain damage.

Chance surrounds our lives with uncertainty. Trying to making sense of this uncertainty leads some to seek solace in religion, others to abandon religion altogether. Realization of the haphazardness of life can lead certain individuals to depression and despair. Such people may see chance as at the root of something undesirable, because it is associated with uncertainty and unpredictability, and hence with danger. Perhaps they still believe that if things are not certain—if the harvests fail and the rains do not come—then there are serious consequences that are readily identifiable with divine punishment.

A more rational reaction to chance is to examine and try to understand a little about how it operates in a variety of different circumstances. To make this possible we first have to be able to quantify uncertainty—in other words, we have to be able to measure it. To do this we use probability. A probability is a number which gives a precise estimate of how certain we are about something; it formalises our usual informal assessment of an event's uncertainty reflected in the use of words such as, *impossible, unlikely, possible*, and *probable*, or a phrase such as *never in a month of Sundays* and *a snowball's chance*. Making the uncertainty more precise may help in many situations.

Nowadays, we encounter probabilities (or their equivalents, such as odds), everywhere we turn: from the money we may occasionally bet on the Derby (Epsom or Kentucky) to the evaluation of a diagnostic test for some disease. Even those arch despisers of uncertainty, physicists, have, during the twentieth century, been forced to use probability theory for studying and attempting to understand the exotic behaviour of subatomic particles. Pure chance is mow considered to lie at the very heart of nature, and one of the greatest triumphs of human intellect during the last century has been the insights gained into how the operation of chance, synonymous as it is with chaos and randomness, leads to so much structure, order and regularity in the Universe.

Chance, risk, probability, statistics and statisticians are the topics to be dealt with in this book. The story will range from the earliest gamblers, who thought that the fall of the dice was controlled by the gods (perhaps it is), to the modern geneticist and quantum theory researcher trying to integrate aspects of probability into their chosen speciality. Subplots will involve tossing coins, rolling dice, coincidences, horse racing, birthdays, DNA profiling, clinical trials and alternative therapies.

Ignoring Stephen Hawking's caution that each equation included in a book will halve its sales, I have given those readers who are interested in more details some very simple mathematics, primarily arithmetic, in several chapters. In most cases this mathematical material is 'boxed-off' so that readers uneasy with mathematics can avoid being distracted from the main story. However, it will often be worthwhile to make a little effort to follow such sections—I wish you luck.

London, August 1999.   B.S. Everitt

Acknowledgments to the First Edition

I am grateful to Littlewoods Pools for information about the development of football pools in the United Kingdom. Likewise I thank William Hill, the Bookmakers for supplying details of the result and prices in the 1995 Epsom Derby. Professor Isaac Marks was kind enough to point out the coin tossing incident in Tom Stoppard's *Rosencrantz and Guildenstern are Dead* and Eric Glover suggested the Holbein painting used in Chapter 1. I would also like thank the many authors from whose work I have benefited in the course of writing this book.

I am grateful to the following for allowing me to reproduce copyrighted material:

(1) Fourth Estate for the extract from *The Man Who Loved Only Numbers* by Paul Hoffman, used in Chapter 10.
(2) Harper Collins for the extract from *Life: An Unauthorized Biography* by Richard Fortey, used in Chapter 15.
(3) Partridge Press and Transworld Publishers Ltd. for the extract from *Lester-The Autobiography* used in Chapter 7.
(4) Faber and Faber for the extract from *Arcadia* by Sir Tom Stoppard used in Chapter 15.
(5) Random House UK Ltd for the extract from *The Inflationary Universe* by Alan Guth, used in Chapter 15.
(6) Perseus Books and Orion Publishing for the extract from *What Mad Pursuit: A Personal View of Scientific Discovery* by Francis Crick, used in Chapter 15.
(7) *The Guardian* for the extract from the article *No Alternative* by Catherine Bennett used in Chapter 13.
(8) *The Independent* and Professor Edzard Ernst for the extract from his article *Complementary Medicine* used in Chapter 13.

My colleagues Dr. Sophia Rabe-Hesketh and Dr. Sabine Landau read the manuscript and made many helpful suggestions, although I was somewhat surprised that they both felt it necessary for the meaning of the initials MCC to be spelt out. My wife Mary-Elizabeth also made encouraging noises when she read the text, as well as undertaking her usual role of correcting the spelling and improving the punctuation. My 6-year-old daughter, Rachel, also tossed a few coins and rolled a few dice once we had agreed on a satisfactory fee. John Kimmel, Jonathan Cobb and Steven Pisano of Springer-Verlag were kind enough to offer much good advice and encouragement, and the efforts of the copyeditor, Connie Day, substantially improved the text. Finally, my secretary, Harriet Meteyard, roamed the Internet for material and in a very real sense acted as collaborator in producing the book.

Contents

A Brief History of Chance

How often things occur by the mearest chance.
Terence, *Phormio*

In the beginning was Chance, and Chance was with God and Chance was God. He was in the beginning with God. All things were made by Chance and without him was not anything made that was made. In Chance was life and the life was the light of men.
Luke Rhinehart, *The Dice Man*

Introduction

Chance plays a central role in many aspects of all our lives. A variety of chance-related terms frequently and effortlessly roll off our tongues. 'How *likely* is it to be sunny this weekend?' 'What are the *odds* of an English player winning a singles title at the next Wimbledon tennis championship?' 'What is the *probability* that my lottery selections will come up this week?' 'What is the *risk* of another Chernobyl happening in the next year?' 'How *probable* is it that I will catch the flu this winter?' But even though such expressions are widely used, are they really understood? How well can the average person evaluate and compare the risks or chances of particular events? There is much evidence that the answer is 'not very well at all', a situation we shall try to improve in the following chapters by looking at various aspects of chance, probability and risk—some simple, some not so simple. To begin, let's consider what we can learn from history.

A Brief History of Chance

Whether people in the Stone Age pondered their chances of dying a violent death as a result of an encounter with a sabre-toothed tiger would be pure speculation, and it is safer, particularly for a nonhistorian, to begin this

B. Everitt, *Chance Rules*, DOI: 10.1007/978-0-387-77415-2_1,
© Springer Science+Business Media, LLC 2008

brief history a little less distant in time. Civilization has always known games of chance. Board games involving chance were perhaps known in Egypt 3000 years before Christ, and the use of chance mechanisms was widespread in pagan antiquity. Palamandes invented games of chance during the Trojan Wars to bolster up the soldiers' morale, and to prevent them suffering from boredom during the lengthy gaps between opportunities to hack members of the opposing army to pieces.

The element of chance needed in these games was at first provided by tossing *astragali*, the ankle bone of the sheep or any other cloven-footed animals; these bones could rest on only four sides, the other two sides being rounded. It appears that the upper side of the bone, broad and slightly convex, counted four; the opposite side, broad and slightly concave, three; the lateral side, flat and narrow, one, and the opposite narrow lateral side which is slightly hollow, six. The numbers two and five were omitted.

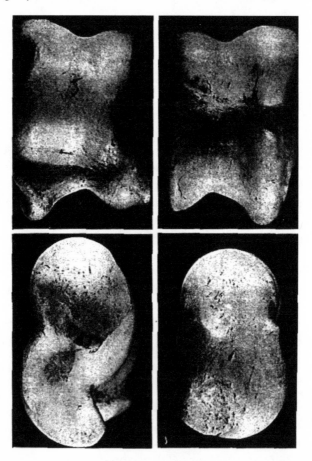

Four views of an astragalus

Egyptian tomb painting showing a nobleman in the after life using an astragalus in a board game

The transition from the astragalus to the cubical die with which we are now familiar probably took place over thousands of years, and it is possible that the first primitive dice were made by rubbing the round sides of the astragulus until they were approximately flat. It seems likely that the earliest die evolved with the Greeks, who called it *tessera* from the Greek for 'four'—the reference is to the four edges of one side of a die. Other early dice made of well-fired buff pottery have been found in Northern Iraq and in India.

But the use of astragali outlived the invention of the die, and they continued to play a prominent part in the gaming that emerged amongst the Romans as a popular recreation for all classes and types of people. The enjoyment of gaming became so widespread that it was perhaps inevitable that some self-appointed guardians of public morality found it necessary to promulgate laws forbidding it, except during the month of December at the Saturnalia. Like most such prohibitions throughout history, they were not very effective. Though repeatedly renewed, they were largely ignored, even by the emperors themselves. Suetonius tells us, for example, that Augustus Caesar

> did not mind being called a gambler; he diced openly, in his old age, too, simply because he enjoyed the game—not only in December when the licence of the Saturnalia justified it, but on other holidays, too, and actually on working days.

In the Jewish community of the time, the rabbis often viewed games of chance with disapproval and gamblers were frequently regarded as thieves. Here, however, what was considered wrong was not the gambling itself, but the fact that the winner took his winnings without paying fair compensation.

Detail from the 'Passion of Christ' by Holbein

But in these early days, chance also had a more insidious role to play. The roll of the dice (or some similar chance outcome) was often the deciding factor in the administration of justice. One can only speculate on how many innocent souls met a horrible death under the wheels of a chariot or provided a meal for a hungry lion simply because an astragalus fell one way rather than another. There were also many times where a chance mechanism, usually the drawing or casting of lots, was used to decide who, among litigants whose rights were equal, should prevail. (The word *lots* is of Germanic origin and found its way into the Romance languages through the Italian, *lottery*, and returned to English by this route.) In the Bible, lots are often used to ensure a fair division of property or privileges as well as to apportion duties and obligations. Perhaps the most infamous instance is provided by Pontius Pilate's soldiers casting lots for Christ's robe as He suffered on the cross. The Bible also offers other examples:

Numbers 25:55. The allocation of the land to the tribes of Israel and its apportionment among individual clans and families. This was commanded to Moses:

The land shall be divided by lot.

Joshua 14:1-2. Carrying out the command given to Moses:

These are the inheritances which the people of Israel received in the land of Canaan, which Eleazer the priest and Joshua, the son of Nun ..... distributed to them. Their inheritance was by lot.

Chronicles 6:39. The order in which the priestly families were to officiate in the Temple:

They organized them by lot.

Judges 20:9-10. Conscripts for military and other duties were chosen by lots. In the punitive action against Gibeah of Benjamin, for example:

We will go up against it by lot and we will take ten men of a hundred.....

(Historically the word lot has a dual meaning. It has meant not only an object that is used to determine a question by a chance mechanism but also one's share of worldly reward determined by divine providence. The latter meaning is more consistent with the use of lots in the Bible, because in biblical times, the outcome of the casting of lots was considered to be the result of divine guidance rather than chance.)

Given that a lot cast to one's advantage might dramatically improve one's life, people must often have been tempted to interfere with the lots in some way so as to increase their own chances of winning whatever prize was at stake (or of avoiding some penalty). It appears that rabbis in particular were deeply aware of the need to neutralize any such possibility, and to this end they imposed various restrictions intended to prevent tampering with the lots. They apparently recognized quite early that a fair procedure is one that, if it is repeated many times, gives each possible outcome with approximately the same frequency. In particular, when there are only two possibilities, the requirement is that in a sufficiently long sequence of events, neither the one nor the other outcome should appear in a significant majority. This is what the rabbis looked for in the casting of lots to ensure that nobody had cheated. But they also used the same frequency argument in other contexts to decide whether events occurred by chance or had to be attributed to some causative agent or law; when most of the events produced the same result, the rabbis concluded that chance could not be the explanation. We will return to these issues in greater detail later.

Apart from its relatively mundane use in deciding the allocation of property or position, the casting of lots (or some equivalent procedure) was, at one time or another in nearly all religions, a fundamental mechanism whereby the deity could be consulted and given an opportunity to make his wishes known to the inquirer. Throughout history, societies have sought information from their own particular god by reading entrails or casting lots. Enthusiasm for this practice was not universal, however. Note Cicero's early scorn, in his essay *De Divinatione*, for the idea of 'divining the future';

> For we do not apply the words 'chance', 'luck', 'accident', or 'causality' except to an event which has not occurred or happened that it either might not have occurred at all, or might have occurred in any other way. How then is it possible to foresee or to predict an event that happens at random, as a result of blind accident, or of unstable chance?

The fondness for communication directly with God (or the gods) by casting lots may explain why in Greece, Rome and other locations, where gambling was common, chance was not studied as a branch of mathematics in the same way as geometry, arithmetic and algebra. Perhaps such study was inhibited by the belief that the operation of chance involved information available only to the gods and that dabbling with random devices was a serious theological business not to be tampered with.

Throwing dice and lots for purposes of divination is still in use today among some remote tribes, but it has long been frowned on by Christians, the Catholic Apostolic Church having condemned sortilege as a relic of paganism. Some Christian sects chose, however, to ignore this condemnation. As late as the eighteenth century, John Wesley sought guidance in this way, giving in his journal, on 4 March 1737, the following charming account of his use of chance to decide whether or not he should marry:

> Having both of us [Mr. Delamotte and himself] sought God by deep consideration, fasting and prayer, in the afternoon we conferred together, but could not come to any decision. We both apprehended Mr. Ingham's objection to be the strongest, the doubt whether she was what she appeared. But this doubt was too hard for us to solve. At length we agreed to appeal to the Searcher of Hearts. I accordingly made three lots. In one was writ 'Marry'; in the second 'Think not of it this year'. After we had prayed to God to 'give a perfect lot', Mr. Delamotte drew the third in which were the words, 'Think of it no more'. Instead of the agony I had reason to expect I was enabled to say cheerfully 'Thy will be done'. We cast lots again to know whether I ought to converse with her anymore, and the direction I received from God was 'Only in the presence of Mr. Delamotte'.

Unfortunately there appears to be no record of what the lady in question thought about this approach to deciding her fate.

A far less agreeable use of lots is recorded to have taken place in the Australian penal settlement on Norfolk Island during the savage punishment regime of James Morriset in the 1830s: 'A group of convicts would choose two men by drawing straws: one to die, the other to kill him. Others would stand by as witnesses'.

After the collapse of Rome in A.D. 455, early Christians more or less rejected the concept of chance. According to St. Augustine, for example, nothing happened by chance, everything being minutely controlled by the will of God. If events appear to occur at random, it is because of human ignorance, not in the nature of the events. Man's true endeavour was to discover and submit himself to the divine will.

Despite such pious hopes, games of chance and gambling flourished and were the main driving force behind a steadily increasing interest in probability until well beyond the Middle Ages. Playing cards, introduced sometime in the fourteenth century, slowly began to displace dice both as instruments of gaming and for fortune-telling. They were, however, too expensive for use by the average medieval gambler, so dice games remained popular for several hundred years. Throughout the centuries, most serious gamblers have developed a firm grasp of the relative chances of drawing the various hands at cards, and of the various outcomes of throwing dice, simply by being intensely involved in and observing many hundreds of games. (The same phenomena can be encountered today by visiting your local betting shop and asking one of the long-term occupants about the likely return on a five-horse accumulator wager.) But slowly gamblers realized that games of chance might become more profitable (or at least less likely to result in penury) if they could involve the mathematicians of the day in calculating the odds involved. Not even the most famous mathematicians were excused, as the following extract from Galileo's *Sopra le Scoperte dei Dadi* (*Thoughts about Dice-Games*) concerning rolling three dice demonstrates:

> The fact that in a dice-game certain numbers are more advantageous than others has a very obvious reason, i.e. that some are more easily and more frequently made than others, which depends on their being able to be made up with more variety of numbers. Thus a 3 and an 18, which are throws which can only be made in one way with three numbers (that is, the latter with 6,6,6, and the former with 1,1,1, and in no other way) are more difficult to make than, for example, 6 or 7, which can be made in several ways, that is a 6 with 1,2,3, and with 2,2,2, and with 1,4,4 and a 7 with 1,1,5; 1,2,4; 1,3,3; 2,2,3,

It seems that Galileo and other Italian mathematicians were responsible for one of the crucial steps in the development of the calculus of chance: recognition that the six faces of a fair die were equally likely to be the result of a throw. During the next two hundred years many other illustrious mathematicians, such as Fermat (of 'last theorem' fame), Pascal (of arithmetic triangle fame) and Huygens (of rings of Saturn fame) built on the work of Galileo with games of chance and gambling remaining their primary motivation. Even in the early seventeenth century, Abraham De Moivre, exiled from France because of religious persecution, sat daily in Slaughter's Coffee House in Long Acre, London, at the beck and call of gamblers, who paid him a small sum for calculating odds. It was not until later in the seventeenth century that James Bernoulli transformed what was then known about the operation of chance from a guide to wagering to a true mathematical theory.

James Bernoulli (1654–1705) was the eldest of four sons, and although his father encouraged him to pursue a career in theology, he turned instead to mathematics, producing in 1681 a mathematical theory of the motion of comets. In 1687, he became professor of mathematics at the University of Basel. In the long tradition of philosophical and theological thought to which James Bernoulli was heir, the idea of probability was not closely tied to the idea of chance. Pascal, Fermat and Huygens did not even use the word probability in their writings on chance. In his work *Ars Conjectandi*, Bernoulli attempted—on the whole successfully—to develop the theory of games of chance into a more formal mathematical theory of probability.

Although cards and dice were as popular in England as anywhere else in Europe in the sixteenth and seventeenth centuries, English mathematicians of the time appear to have made little contribution to the theory of probabilities. They may have all disapproved of gambling but perhaps it is more likely that they realized there was little possibility of

JAKOB BERNOULLI um 1687

James Bernoulli

profit in discussing the fall of the dice. But although he was not involved in developing the more theoretical aspects of probability, it was an Englishman, John Graunt, who extended the area of application of empirically derived probabilities to outside the gaming tables of the day.

John Graunt (1620–1674) was the son of a London draper, and after serving an apprenticeship, he joined his father's business, which he eventually took over. He received an ordinary education in English learning and studied Latin and French on his own in the morning before business hours. He soon became a respected London citizen and held a number of important offices in the Draper's Company, in the Ward offices, and in the council of the city. He was a captain and later became a major in the trained band of the city. He also had many acquaintances in cultural and scientific circles. Graunt was what might be termed a vital statistician; he examined the risk inherent in the processes of birth, marriage and death and used bills of mortality to compare one disease with another and one year with another by calculating mortality statistics—that is, the chances of dying. An example is his work on determining which had been the worst plague year by comparing all burials with the number of burials involving plague victims. See the accompanying table.

Period	All burials	Plague victims	% Plague
1592	26,490	11,503	43
1593	17,844	10,662	60
1603	42,042	36,269	86
1625	54,265	35,417	65
1636	23,359	10,400	45
1665	97,306	68,396	70

The last column contains what are essentially empirically derived relative frequency probabilities (see Chapter 2) of death due to plague in a particular year.

Huygens, Nicholas Bernouilli, Halley and De Witt built upon Graunt's work, and their investigations eventually led to the development of the mathematics behind what actuaries and insurance companies still use to this day—quantification of the chance of dying at particular ages from specific causes.

During the eighteenth and nineteenth centuries, the techniques of modern mathematical statistics began to be laid down—all on the foundations of earlier work on probability and chance. In the twentieth century, statistics has become known as the mathematical tool for analyzing experimental and observational data. Enshrined by public policy as the only reliable basis for judgements as to the efficacy of medical procedures or the safety of chemicals, and adopted by business for such uses as industrial quality control, statistics is among the products of science whose influence on public and private life has been most pervasive. (We shall try to demonstrate this point in Chapter 12 using the particular example of medical research.) But despite its current respectability, it should be remembered that the origins of modern statistics lie in the inns and coffee houses of fifteenth- and sixteenth-century Europe, where inveterate gamblers desperately tried to increase their chances of making a successful bet.

The twentieth century has also seen chance being accepted as a fundamental feature of the Universe, as physicists have asserted the probabilistic nature of matter at the particle level. It is a basic postulate of quantum theory, the fundamental theory of matter, that individual events at the subatomic level are neither reproducible at will or by experiment nor predictable in theory. We can speak only of the probability of such events. The concept of probability represents the connection between the irreproducible, unpredictable, single event and the predictability and uniformity of multiple events. Erwin Schrödinger referred to this as the principle of order from disorder, although he was not sympathetic to the introduction of probability theory into his beloved physics. 'God knows I am no friend of probability theory', he told Einstein. 'I have hated it from the first moment that our dear friend Max Born gave it birth.' Schrödinger, like most physicists of the time, still longed for certainty.

But for Max Born chance was an even more fundamental concept than causality, since he considered that whether or not a cause–effect relation holds can only be judged by applying the laws of chance to the relevant observations. But what are the laws of chance and how do we assign probabilities to chance events? Cue Chapter 2.

What Are the Chances? Assigning Probabilities

Chance, too, which seems to rush along with slack reins, is bridled and governed by laws.
Boethius, *The Consolation of Philosophy*

Misunderstanding of probability may be the greatest of all impediments to scientific literacy.
Stephen Jay Gould

Introduction

Probability plays a central role in quantifying chance with the probability of an event being a measure of the event's uncertainty. When people talk about the probability of an event, say, the probability of rain tomorrow, they may have in their mind a scale of expressions that distinguish the different degrees of probability in some way, for example, 'it is certain not to rain', 'it is very unlikely to rain', 'it is unlikely to rain', 'it is as likely to rain as not', 'it is likely to rain', 'it is very likely to rain', 'it is certain to rain'. But these expressions, apart from the first, fourth and last are vague and different people may interpret them differently. The vagueness is of little consequence when discussing the weather with a friend but might be of concern in conversations with your bookmaker, insurance salesman or doctor. We need a more precise, *numerical* scale for probability.

By convention, probabilities are measured on a scale from zero to one. To say an event with probability of one fifth (1/5 or 0.2) implies that there is a 1 in 5 or 20% chance of the event happening. The zero point on the probability scale corresponds to an event which is impossible—the author now ever running a marathon in less than two and a half hours suggests itself. (Come to think of it, that was always pretty unlikely!) The highest point on the probability scale, one, corresponds to an event which is certain to happen—the author's death within the next 35 years unfortunately comes to mind. But how do we assign probabilities to events that are

B. Everitt, *Chance Rules*, DOI: 10.1007/978-0-387-77415-2_2,
© Springer Science+Business Media, LLC 2008

uncertain but possible, rather than certain or impossible? Well, there are a number of possibilities.

Using Relative Frequencies as Probabilities

Suppose 1,000,000 children are born in a particular year with 501,000 of them being boys. The relative frequency of boys is simply 501,000/ 1,000,000 = 0.501. Now suppose that in a small village somewhere, 10 children are born in the year, of whom 7 are boys. The relative frequency of boys in this case is 7/10 = 0.7. As the rabbis of over two millennia realized, the relative frequency of an event relates to the event's chance or probability of occurrence. In fact, this concept is now used to provide an intuitive definition of the probability of an event as the relative frequency of the event in a *large* number of trials. Thus in the examples given, the observation of 501,000 male births amongst a million births suggests we can claim that the probability of a male birth is 0.501, but it would be unconvincing to use the counts of births from the small village to suggest that the probability of a male child is 0.7.

The obvious question that needs to be asked here is, 'How large is large enough?' Sadly there is no simple answer, which may appear to imply that the relative frequency approach to assigning probabilities is not perfect. But before jumping to such a conclusion we might ponder for a moment that a perfect model of the solar system was not needed to successfully send probes to photograph the outer planets. And relative frequency probabilities remain very useful despite being somewhat less than perfect.

The Classical Definition of Probability

Suppose we are interested in assigning a probability to the event that a fair coin, when tossed, comes down heads. We could use the relative frequency approach, by simply tossing the coin a large number of times and then dividing the number of heads obtained by the total number of tosses. Undertaking this task myself I tossed a 50-pence (50p) coin 500 times and obtained 238 heads, leading to a relative-frequency-based value for the probability of a head of 0.48. But unlike this frequency approach, the so-called classical definition allows probabilities to be assigned to single events. We might, for example, be interested in finding the probability of getting a two on a single roll of a die. Calling the case of two 'favourable', the numerical value of its probability is found by dividing the number of favourable possible cases by the total number of all equally possible cases. For a true die there are six outcomes, all of which are equally possible, but only one of which is favourable; consequently the probability of throwing a two is 1/6.

In essence, this approach to assigning probabilities assumes that without further information about the physical process that, say, flips a coin or rolls a die, a sensible observer would view all outcomes as being equally likely; hence one half for the probability of a head and one sixth for the probability of each side of a die.

We shall look more at tossing coins in Chapter 4 and at rolling dice in Chapter 5; here we will consider a different example of assigning probabilities using the classical approach.

Let's look at the following three problems involving leap years;

1. What is the chance of a year, which is not a leap year, having 53 Sundays?
2. What is the chance that a leap year, will contain 53 Sundays?
3. What is the chance that a leap year which is known not to be the last year in the century should be a leap year?

To get the answer to each question we need to count the 'favourable' outcomes in each case and divide the resulting number by the total number of equally possible outcomes. In this way we are led to the following solutions;

1. A non-leap year of 365 days consists of 52 complete weeks, and one day over. This odd day may be one of the 7 days of the week, and there is nothing to make one more likely than another. Only one will lead to the result that the year will have 53 Sundays; consequently the probability that the year has 53 Sundays is simply 1/7.
2. A leap year of 366 days consists of 52 complete weeks, and 2 days over. These days may be

   Sunday and Monday,
   Monday and Tuesday,
   Tuesday and Wednesday,
   Wednesday and Thursday,
   Thursday and Friday,
   Friday and Saturday,
   Saturday and Sunday.

   And all seven possibilities are equally likely. Two of them (the first and the last) will produce the required result so the chance of a leap year containing 53 Sundays is 2/7, twice the probability for a non-leap year.

3. The year may be any of the remaining 99 of any century, and all these are equally likely; 24 of the years are leap years so the chance that the year in question is a leap year is 24/99.

Subjective Probability

The classical and relative frequency methods of assigning probabilities cannot be used in all circumstances where we are interested in making probability statements. What, for example, is the probability that life exists on a planet circling the nearest star, Proxima Centauri? For that matter, what is the probability that there even *is* a planet circling Proxima Centauri? The classical definition of probability provides no answer for either question, and there are no relative frequency data available that might help. In such circumstances we are forced to express a subjective opinion about the required probabilities—to assign a subjective probability. We might use experience, intuition, or a hunch to arrive at a value, but the assigned probability will be subjective, and different people can be expected to assign different probabilities to the same event. Subjective probabilities express a person's degree of belief in the event and so are often referred to as personal probabilities.

A useful short summary of subjective probability has been provided by Peter Lenk of the University of Michigan:

> In the frequency interpretation, probability is an innate characteristic; flipped coins come up heads 50% of the time because that is the property of coins. In subjective probability, flipped coins come up heads 50% of the time because the observer does not have information that would lead him or her to believe that one side is preferred over the other. The locus of uncertainty is not the coin and its environment, but resides within the observer. If the observer knew more about the physical laws of coin flips and the initial conditions, then he or she could refine his or her probability assessment.

Although we shall not be concerned with subjective probabilities in the remainder of this book one word of advice may be appropriate: Never let the values of the subjective probabilities you assign be zero or one. For example, my own subjective probabilities for flying saucers and the Loch Ness monster are very near to zero, but there no doubt exist people whose subjective probabilities for these two phenomena are close to one. And my own current very low subjective probability for flying saucers, say, would increase dramatically if such a machine landed in my garden.

For scientists in particular, complete disbelief (a subjective probability of zero) or complete certainty (a subjective probability of one), are equally undesirable. Scientists (and others) need to avoid the paralysis that can be caused by the pursuit of certainty. However successful and reliable a theory, say, may be up to any point in time, further observations may come along and show a need for adjustment of the theory, while at the other extreme, however little confidence one has in a theory, new information may change the situation. But scientists and non-scientists both, court the danger of descending into an emotional pit of pride and prejudice once their subjective probabilities for any event or phenomenon take either the value one or zero. In the words of that well known Chinese philosopher, Chan: 'Human mind like parachute: work best when open.'

Rules for Combining Probabilities

Suppose we roll a fair die and want to know the probability of getting either a six or a one. Since a six and a one cannot occur together they are usually called *mutually exclusive events*, and the probability of either of them occurring is simply the sum of the individual probabilities, in this case $1/6+1/6 = 1/3$. This is a general rule for finding the probability of mutually exclusive events; if there are two such events A and B then

$$Pr(A \text{ or } B) = Pr(A) + Pr(B),$$

where Pr stands for 'the probability of' and the equation is translated as 'the probability of event A or event B is equal to the probability of event A plus the probability of event B'. So the die example is described by

$$Pr(6 \text{ or } 1) = Pr(6) + Pr(1).$$

For three mutually exclusive events, A, B and C,

$$Pr(A \text{ or } B \text{ or } C) = Pr(A) + Pr(B) + Pr(C).$$

For example, tossing an even number with a fair die,

$$Pr(2 \text{ or } 4 \text{ or } 6) = Pr(2) + Pr(4) + Pr(6) = 1/6 + 1/6 + 1/6 = 1/2,$$

and so on for more than three mutually exclusive events.

Suppose now we roll two dice separately and want the probability of getting a double six. Since what happens to each die is not affected by what happens to the other, the two events, six with one die and six with the other, are said to be *independent*. The probability that both occur is then simply obtained by multiplying together the probabilities of each event, i.e., $1/6 \times 1/6 = 1/36$. This is a general rule for finding the probability of independent events; if there are two such events, A and B then,

$$\Pr(A \text{ and } B) = \Pr(A) \times \Pr(B).$$

For three independent events, A, B and C,

$$\Pr(A \text{ and } B \text{ and } C) = \Pr(A) \times \Pr(B) \times \Pr(C).$$

For example getting a triple six when rolling three dice,

$$\Pr(\text{triple six}) = \Pr(6) \times \Pr(6) \times \Pr(6) = 1/6 \times 1/6 \times 1/6 = 1/216$$

and so on for more than three independent events.

By combining the *addition rule* and the *multiplication rule*, probabilities of more complex events can be determined, as we shall find in later chapters.

Odds

Gamblers usually prefer to quantify their uncertainty about an event in terms of *odds* rather than probabilities, although the two are actually completely synonymous. An event with a probability of 1/5 would be said by an experienced gambler to have odds against of 4 to 1. This simply expresses the fact that the probability of the event not occurring is four times that of it occurring. Similarly, an event with probability 4/5 would be said by the gambler, to be 4 to 1 on—here the probability of the event occurring is four times that of it not occurring. The exact relation between odds and probability can be defined mathematically:

- An event with odds of 'F to 1 in its favour (odds-on)' has probability $F/(F+1)$.
- An event with odds of 'A to 1 against' has probability $1/(A+1)$.
- An event with probability P implies that the odds in favour are $P/(1-P)$ to 1, whereas the odds against are $1/P-1$.

Odds and probabilities are completely equivalent ways of expressing the uncertainty of an event but gamblers generally deal in the former and scientists in the latter, although it has to be said that many medical statisticians are, perhaps due to a misspent youth, also very adept with using odds.

Choice and Chance; Permutations and Combinations

Life seems to be a choice between two wrong answers.
Sharyn McCrumb

The permutations and combinations are endless. It's like a game of three-dimensional chess.
Sherry Bebtich

*I failed math twice, never fully grasping probability theory.
I mean, first off, who cares if you pick a black ball or a white ball out of the bag? And second if you're bent over about the color, don't leave it to chance. Look in the damn bag and pick the color you want.*
Stephanie Plum, *Hard Eight*

Introduction

We have continually to make a choice between different courses of action open to us and our future happiness may depend on the choice we make. Sadly, arithmetic is of no use in guiding our choice. Here arithmetic has to give way to some higher philosophy. But arithmetic as the science of counting and calculation does have something to tell us about the *number* of ways we can exercise our choice. For example, when we go to vote, we may have to vote for, say, two candidates out of four who are standing for election. Arithmetic has nothing to do with the manner in which we exercise our privilege as a voter but it can determine the number of ways two candidates can be selected from the four available. The arithmetic that enables us to determine the number of choices in particular situations is a stepping stone to assigning probabilities in these situations via the classical approach outlined in the previous chapter. The arithmetic involved is that of permutations and combinations.

B. Everitt, *Chance Rules*, DOI: 10.1007/978-0-387-77415-2_3,
© Springer Science+Business Media, LLC 2008

Permutations and Combinations

Sefer Yetsirah or *Book of Creation* is a Hebrew text written sometime between the second and eighth centuries A.D. For its unknown author, the twenty-two letters of the Hebrew alphabet represent the building blocks of creation; '[God] drew them, hewed them, and combined them, weighed them, interchanged them, and through them produced all of creation and everything that is destined to be created.'

Sefer Yetsirah contains less than sixteen hundred words and much of it is devoted to enumerating various arrangements of the letters of the alphabet. For example, the number of arrangements of the twenty-two letters taken two at a time is given as follows:

> Twenty-two element-letters are set in a cycle in two hundred and thirty-one gates—and the cycle turns forwards and backwards... How is that?...Combine A with the others, the others with A, B with the others, and the others with B, until the cycle is completed.

Here *forwards* and *backwards* refer to the fact that in 231 arrangements the letters appear in the order of the alphabet, the earlier one in the alphabet followed by the later one, while in the 'backwards' 231 arrangement the order is reversed—'A with the others', means that with A in the first place there are twenty-one pairs, and similarly for every one of the twenty-two letters of the alphabet, so the total number of arrangements is $22 \times 21 = 462$.

Along the same lines and at about the same time, Donnolo gives the following proof for the rule that n letters can be arranged in $n \times (n - 1) \times (n - 2) \times \ldots .3 \times 2 \times 1$ ways (this product of the numbers from n down to 1 is now written in a short-hand form as $n!$—read as factorial n—so, for example, 6! is 720).

> A single letter stands alone but does not form a word. Two form a word: the one proceeding the other and vice versa—give two words, for twice one is two. Three letters form three times two that is six. Four letters form four times six that is twenty-four...and in this way continue for more letters as far as you can count. The first letter of a two-letter word can be interchanged twice, and for each initial letter of a three- letter word the other letters can be interchanged to form two-letter words—for each of three times. And all the arrangements there are of three-letter words correspond to each one of the four letters that can be placed first in a four letter word: a three-letter word can be formed in six ways, and so for every initial letter of a four-letter word there are six ways —altogether four times six making twenty-four words...and so on.

Sefer Yetsirah inspired generations of mystics to speculate on more complicated problems, such as determining the number of arrangements of letters some of which are identical. Fascinating though such speculations may be to Hebrew scholars and academics, why, in a book whose main subject is chance, are *we* spending time on the topic? The answer is that in many situations, it is intimately connected to the enumeration of probabilities via what we would nowadays call *combinations* and *permutations*. Briefly examining both will greatly assist us in later discussions.

First, we need to get a clear idea of the difference between the terms, combination and permutation. In fact, the difference is simply one of whether *order* is important. Consider the five letters of the name

$$\text{SMITH}$$

Suppose we are interested in permutations and combinations of, say, letters from the name. The arrangements

$$\text{SMI, SIM , MSI, MIS, ISM, IMS}$$

count as different permutations of the three letters. But they represent a single combination of three letters, because combinations disregard order. The number of combinations of three letters taken from the five can be shown to be

$$5!/(3! \times 2!) = 10.$$

(This is an example of the general formula for the number of ways of selecting r objects from n objects without regard to their order namely,

$$n!/[r! \times (n - r)!].$$

Numbers of this kind are known as *binomial coefficients*.)

The ten combinations of the three letters are SMI, SMT, SMH, SIT, SIH, STH, MIT, MIH, MTH, ITH. The three letters chosen in each combination can be rearranged in $3! = 6$ different orders, so the number of permutations is $6 \times 10 = 60$.

For four letters the corresponding results are

$$\text{Number of combinations} = 5!/(4! \times 1!) = 5,$$

and number of permutations $= 4! \times 5 = 120$. For five letters, there is only a single combination and the number of permutations is $5! = 120$.

Using Permutations and Combinations to Assign Probabilities

Let's now look at a few examples of the use of permutations and combinations in assigning probabilities since a little experience gained here will be very helpful in later chapters.

Example 1
The four letters *s*, *e*, *n* and *t* are placed in a row at random; what is the chance of their standing in such order as to form an English word?
The four letters can be arranged in 4!=24 different orders (permutations of the four letters); all are equally likely. Four of the arrangements produce an English word,

$$sent, \ nest, \ nets, \ tens.$$

So the required probability is just $4/24 = 1/6$.

Example 2
A bag contains five white and four black balls. If they are drawn out one by one, what is the chance that the first will be white, the second black, and so on alternately?

Assuming the white balls are all alike and the black balls are all alike, the number of possible combinations of ways in which the nine balls can be drawn from the bag is

$$9!/(4! \times 5!) = 126.$$

The balls are equally likely to be drawn in any of these ways; only one of the 126 possibilities is the alternate order required, so the probability of white, followed by black, followed by white etc., is 1/126.

Example 3
Four cards are drawn from a pack of 52; what is the chance that there will be one card of each suit?

Four cards can be selected from the pack in $52!/(4! \times 48!) = 270725$ ways. But four cards can be selected so as to be one of each suit in only $13 \times 13 \times 13 \times 13 = 28561$ ways. So the required probability is 28651/270725 which is just a little over 1/10.

And finally one problem to leave for my reader(s):

- In how many ways can an arrangement of four letters be made out of the letters of the words *choice and chance*?

Answers on a postcard please (or these days, email perhaps).

Tossing Coins and Having Babies

4

> *The equanimity of your average tosser of coins depends upon the law, or rather a tendency, or let us say a probability, or at any rate a mathematically calculable chance, which ensures that he will not upset himself by losing too much nor upset his opponent by winning too often.*
> Guildenstern *in* Tom Stoppard's *Rosencrantz and Guildenstern are Dead*

> *Literature is mostly about having sex and not much about having children. Life is the other way round.*
> David Lodge, *The British Museum is Falling Down*

The Toss of a Coin

Test match cricket played over 5 days is without doubt one of the best arguments in favour of a thoughtful and beneficent God. As a game to be involved in or to watch, it is without equal. For 6 hours a day (weather permitting) the two teams are locked in a struggle in which a single mistake by a batsman can be punished by their dismissal from the field of play, or a brief lapse in concentration by a fieldsman can mean banishment to the furthest reaches of the ground to suffer the taunts of the crowd for several hours. Fast bowlers strain every sinew to extract opposition batsmen by sheer pace or fear, and the spinners follow to examine the remaining batsmen in rather more cerebral fashion. And all the while the crowd can choose to concentrate on the cricket or read the paper, snooze and generally think all is right with the world.

But in many cases the titanic battle of a test match is not decided by the combined skills of the eleven members of a team, by a brilliant individual performance, by pre-match preparation, or the tactics of the captain. Instead the match is frequently decided by the chance event with which it begins, the toss of a coin. The captain who correctly guesses which face the coin will show,

B. Everitt, *Chance Rules*, DOI: 10.1007/978-0-387-77415-2_4,
© Springer Science+Business Media, LLC 2008

and so wins the toss, can choose which side will bat first. In many games this decision will have a tremendous influence on the final result, and my old uncle's advice 'be lucky', if it could be relied upon in practice, might be the optimal criterion on which test selectors should base their choice of a captain.

Bradman and Hammond spinning a coin at the beginning of the Oval Test Match of 1938. Bradman lost and England chose to bat first. Between the two captains is Bosser Martin, the Oval groundsman

There are, of course, occasions where the captain who wins the toss gets it all dreadfully wrong. Such a case was the first match in the 1996–1997 Ashes series (test matches between England and Australia). The toss was won by the Australian captain, Mark Taylor, and he chose to bat first (the choice in the majority of test matches). Australia was 96 for 8 by lunchtime on the first day and eventually made 118. England then scored 478 in their first innings and eventually won the match by 9 wickets.

(I feel it may be necessary to offer an apology here—and, of course, my heartfelt sympathy—to readers in countries where cricket in general, and test match cricket in particular, have yet to make an impact, for a digression about matters that may be doomed forever to be a mystery. On the other hand, perhaps they will be in my debt for exposing them, however briefly, to a pursuit that is to baseball and soccer, say, as Pink Floyd is to the Spice Girls.)

The Probability of a Head

When a coin is tossed there are two exclusive possibilities, each of which has the same uncertainty (assuming that the person making the toss has not learnt to make the coin stand on its edge). Each face has the same chance of occurring on each toss and by the classical definition of probability, we can assign the probability 0.5 equally to the occurrence of a head or that of a tail. How does this compare with the probabilities of a head or tail derived from the relative frequency approach? In the interest of science the author undertook to toss an ordinary 50p coin, 25 times, 50 times and finally 100 times, recording the number of heads that appeared. (The last four rows of this table are not a result of days of tossing a 50p coin but arise from using a computer to simulate the coin tossing experiment.)

Number of tosses	Number of heads	Proportion of heads
25	14	0.5600
50	27	0.5400
100	54	0.5400
500	238	0.4760
1000	490	0.4900
10,000	4958	0.4958
100,000	50,214	0.5021

When we reach large numbers, as the number of tosses increases, the proportion of heads gets closer and closer to 0.5 although it doesn't ever equal 0.5. Mathematically, this slight glitch is overcome by defining the probability of a head as the relative frequency of a head when the number of tosses gets very, very, very large (in mathematical parlance 'goes to infinity'). This necessity for very long sequences of tosses to make the empirically derived probability of a head equal to the value derived from the classical approach is the reason why in, say, 20 tosses of a fair coin, the number of heads is unlikely to be exactly 10. (In fact an exact fifty-fifty split of heads and tails has a probability of a little less than one in five—see the next section for more details.) The particular sequence of heads and tails found in the 20 tosses is unpredictable—the term used by statisticians is *random*. Randomness is synonymous with uncertainty. The sequence of heads and tails will change each time we repeat the process. Ten such sequences involving 20 tosses obtained by the author are the following:

H T T H H H H H T T T H T H T H H H H T—number of heads 12,
H H T H T T T T T H H T T T H H H T T H—number of heads 9,

H T H T H H H H T H T H H T T T T T T T—number of heads 9,
T H H H T H T T H T H T H H H H H H H H—number of heads 14,
H T T H T H H T T H H H T H H T T H H H—number of heads 12,
T T H T H T H H H H H T T T H H H H T H—number of heads 12,
H H H T H T T H H T T H T H H T T T T T—number of heads 9,
H H T T H H T H T H H T H T H T T T T H—number of heads 10,
T T T T H H T T T T H T H T T T T H H T—number of heads 6,
H H H T H T H T H T T T T T T T T H H H—number of heads 9.

Not only are the sequences themselves random (although, see below) but so are particular characteristics of them. The number of heads, for example, varies from 6 to 14. The number of heads in any set of 20 tosses is unpredictable—it is subject to random variation. Such variation has important implications in many areas, as we shall see later, particularly in Chapter 12. But although the number of heads obtained in any set of 20 tosses of a fair coin is unpredictable, we *can* say something about what to expect in the long run. On average we would expect 10 heads (the average number in the ten sequences above is 10.1), with the majority of sequences having between 4 and 16 heads. Values less than 4 or greater than 16 might lead us to question just how fair the coin really is.

(My apologies for launching into a small digression at this point. To mathematicians and scientists a random sequence is one that is so irregular that no way can be found to express it in shorter form. The sequences above generated by my coin tossing were said to be random but what if I had tossed 20 heads? Could this sequence still be labelled as random? Not really, and that's the rub as somebody once said. A chance process such as tossing a coin a large number of times will lead to a random sequence in many cases, but not all. Some statisticians have suggested that sequences generated by a chance process such as tossing a coin should be called stochastic, leaving random for incompressible sequences. But this introduces a complication which I feel is unnecessary in this book and so I shall continue to use random in its more general sense. Digression over.)

Knowing that the probability of getting a head when tossing a fair coin is 0.5 does not help us in making an accurate prediction of whether we will get a head or tail when next we spin a coin. It does, however, say something about the number of heads that are likely to occur in the long run. Just as Maynard Keynes could, with some confidence, predict that 'In the long run we're all dead', knowing the probability of an event allows us to be precise about the event's long-run behaviour.

Suppose we now use the tosses of a single coin to play a simple game. The game involves two players each tossing the coin in turn. The winner is the first player to throw a head. Although each player has an equal chance of getting a head at each toss, do both players have an equal chance of winning the game? Of course not, because the player who spins first has an advantage.

Let's try to determine how much of an advantage. To add a sense of realism to the situation let's call the players Peter and Gordon, and let's assume that Peter gets to go first.

Peter will win if any of the following happen:

(1) Peter gets a head on his first toss—probability, $\frac{1}{2}$.
(2) Peter gets a tail on the first toss, Gordon gets a tail on the second toss and Peter gets a head on the third toss—probability, $\frac{1}{2} \times \frac{1}{2} \times \frac{1}{2}$.
(3) Peter gets a tail on the first toss, Gordon gets a tail on the second toss, Peter gets a tail on the third toss, Gordon gets a tail on the fourth toss, and Peter gets a head on the fifth toss—probability, $\frac{1}{2} \times \frac{1}{2} \times \frac{1}{2} \times \frac{1}{2} \times \frac{1}{2}$
(4) Etc., etc., etc.......

The overall probability of Peter winning is simply the sum of the probabilities of each of these possible situations—that is,

$$\frac{1}{2} + \frac{1}{2} \times \frac{1}{2} \times \frac{1}{2} + \frac{1}{2} \times \frac{1}{2} \times \frac{1}{2} \times \frac{1}{2} \times \frac{1}{2} + \cdots$$

As it stands, this result is neither very attractive nor very informative since it involves a series containing an *infinite* number of terms. Fortunately, mathematicians know how to find the sum of such a series. It is given by

$$\frac{\frac{1}{2}}{1 - \frac{1}{2} \times \frac{1}{2}} = \frac{2}{3}$$

Because either Peter or Gordon must eventually win, the probability that Gordon wins is simply one minus the probability that Peter wins—that is, $1 - \frac{2}{3} = \frac{1}{3}$.

Although each player has an equal chance of getting a head on each toss, the player who goes first has, overall, twice the chance of winning the game compared to the player who goes second. (Starting in other, more complex games such as chess may also bestow a distinct advantage, but one that is far more difficult to quantify.)

The advantage of the player going first in the simple coin-tossing game can be demonstrated numerically by repeatedly playing the game; in 100 games played by my wife and myself, the player going first won 69 times and the player going second had only 31 wins. The maximum length of any one game was six tosses, which brings us to another question: On average, how long will this simple game last? In our 100 games the average number of tosses was 1.83 (I know this cannot occur in practice—it's like the average family size being 2.2. But the average does reflect the fact that most games will be only one or two tosses in length). It can be shown mathematically that this game can be expected to last two tosses and also that games of more than six or seven tosses are relatively rare.

Some Surprising Results About Coin Tossing

Even simple coin-tossing games can give results that most people find surprising. First, let's consider the probability of getting an equal number of head and tails when we toss a coin an even number of tosses. Many people might be tempted to suggest that as the number of tosses increases this probability also increases. But this is not the case; in fact as the number of tosses grows the probability of an equal number of heads and tails goes down. To demonstrate that this is the case we shall consider initially just the case of two tosses and four tosses;

Two tosses
Here the four possible results are HH, HT, TH, TT; all four outcomes are equally likely and two result in an equal number of heads and tails so the required probability is $1/2 = 128/256$ (the reason for expressing the result in this way will soon become clear).

Four tosses
Here the 16 possible results are HHHH, HHHT, HHTH, HTHH, THHH, HHTT, HTHT, HTTH, THHT, THTH, TTHH, HTTT, THTT, TTHT, TTTH, TTTT; all 16 outcomes are equally likely and six result in an equal number of heads and tails so the required probability is $6/16 = 96/256$, just under a half.

For six tosses the probability of three heads and three tails is $5/16 = 80/256$, for eight tosses the equal division of heads and tails probability is $35/128 = 70/256$. And for ten tosses the probability is $63/256$. The probability is seen to be decreasing as the number of tosses increases.

Next suppose we have n identical coins for each of which the probability of a head is $1/2$. We toss all the coins. Those that show heads we put aside and toss again all those that show tails. We then repeat the process until eventually all coins show heads. What is the probability that the number of coins used in the last toss is one and how does this depend on the original number of coins, n? Trying this out myself, tossing five coins ten times, I found the following results for the number of coins used on the last toss:

1 coin, 1 coin, 2 coins, 1 coin, 2 coins, 2 coins, 3 coins, 1 coin, 1 coin, 1 coin.

So, in this experiment, using the relative frequency approach, the estimated probability of there being one coin on the last toss is 0.6. Clearly 10 trials is not enough; the experiment needs to be repeated for a larger number of trials and for different numbers of coins to get accurate empirical probabilities. But, before you spend several happy hours tossing lots and lots of coins, I can tell you that mathematically the probability of the last toss involving a single coin is very close to 0.72 for *any* value of n.

Now consider how many tosses are needed on average to get the first occurrence of a particular pattern (known as the *waiting time* for the pattern); for example, when tossing a coin we might want to know the average number of tosses that would be needed before we get a HH pattern in the sequence of heads and tails. And is this average for the HH pattern different for say an HT or TH pattern? Intuition may tell us the averages are the same; after all each pattern occurs with a probability of one quarter so why shouldn't the waiting time for the pattern to occur also be the same? To begin, we can see if some actual coin tossing suggests whether intuition is correct or not. Below are two sets of results from my tossing a coin; in the first I toss the coin until the pattern HH occurs and in the second until the pattern TH appears, recording in each case the number of tosses until the required pattern is seen, and repeating the exercise ten times.

First coin tossing results, until HH appears:
1. THH (waiting time 3) 2. HH (2) 3. TTHTHTHH (8) 4. TTHH (4)
5. TTHTHTHH (8) 6. HTTTHTHH (8) 7. HTHTHTTHTHTHH (13)
8. HTHTTHTHH (9) 9. TTHTHTHH (8) 10. TTTHH (5)
Average waiting time is 6.8 tosses.

Second coin tossing results, until TH appears:
1. TTH (waiting time 3) 2. HTTTH (5) 3. HTTTTTH (7) 4. HHHTTH (6)
5. HHHHHHTTH (9) 6. TTH (3) 7. TTTH (4) 8. HHTH (4) 9. HHTH (4)
10. HTTTTH (6)
Average waiting time is 5.1.

Based on just these ten sequences it seems that the average waiting time for a TH pattern may be shorter than that for an HH pattern. But sceptical readers might rush to demand many more sequences of tosses—they are right to do so but I will leave that for them to undertake, confident that they will find that the waiting time for the pattern TH is indeed shorter than that for HH. I am confident simply because I know that some clever mathematics shows that for TH the average waiting time is four tosses but for HH it is six tosses. (Before trying to impress your friends with this result remember that the average waiting time can vary quite a bit so you will probably need many repeat sequences before a clear difference in the average waiting times is seen—but it is there.)

The partial explanation for this seemingly odd difference in waiting times for the HH and TH patterns is that when we look at blocks of two tosses there is overlap, so that outcomes are not independent. A full explanation of the different values of the waiting times requires some technical effort and so cannot be pursued here. But just to whet your interest a little more have a look at some other average waiting times for patterns of length three in the repeated tossing of a fair coin;

Pattern Average waiting time (number of tosses): HHH 14, HHT 8, HTH 10, HTT 8.

Finally, as the pattern length increases, the difference in the average waiting time for the all heads pattern and other patterns, becomes greater. For example, for patterns of length six the average waiting time for the pattern HHTTHH is 70 tosses, whereas for HHHHHH it is 126 tosses. Intuition, and its cousin, common sense, are, it seems, not all they are often cracked up to be.

(For those readers interested in the technical details behind the results given above I recommend the article *Coincidences and Patterns in Sequences* by Anirban Dasgupta in the *Encyclopedia of Statistical Sciences* published by Wiley.)

Suppose now we toss a fair coin a large number of times. Each time the coin comes down heads Peter is given a gold coin, and each time it comes down tails Gordon receives the same reward. Most people predict that during the course of the game Peter and Gordon will have about the same amount of money, with Peter in front about half the time and Gordon in the lead also for about half the time. But in practice this is just what *doesn't* happen—one of the players will be in the lead for most of the time! Changes of lead are very rare. (Just to make things completely clear, a change of lead implies that the number of heads and tails first becomes equal, and then at the next toss the lead is taken by the player who was behind at the toss before equality was achieved.) Those readers with time on their hands can show this by indulging in coin tossing for a few hours. But the effect can be demonstrated more simply and less exhaustingly by using a computer to simulate the game. The results of 2000 tosses are shown in the diagram below. We see that between about toss 400 and toss 2000, Peter is always in the lead. This seemingly paradoxical situation can be illustrated more dramatically by imagining two frenetic coin tossers spinning a coin once a second for 365 days. In about 10% of such year-long games, the more fortunate player will be in the lead for more than 362 days! In fact in this game, and for any number of tosses, it is more likely that the lead *never* changes hand than any other specific number of changes. And the probabilities of increasing number of changes of lead decrease steadily from the maximum value for zero changes of lead.

Coin tossing can be used to describe what is often know as the 'Gambler's Fallacy.' Suppose we toss a coin and get 10 successive heads. Many

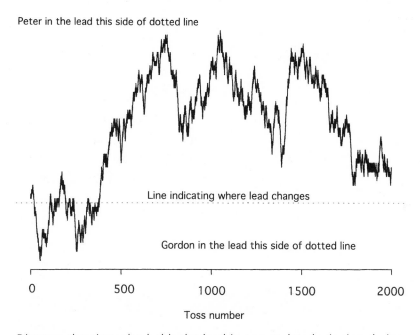

Peter in the lead this side of dotted line

Line indicating where lead changes

Gordon in the lead this side of dotted line

0 500 1 000 1 500 2000

Toss number

Diagram showing who holds the lead in accumulated winning during 2000 tosses of a fair coin

gamblers would be willing to bet that tails will appear next. They argue that tails is more likely to turn up than heads because the 'law of averages' suggests that this will help to even things out. It's this belief that 'my luck's got to change', based on a misconception of the fairness of the operation of chance, that leads so many gamblers astray. Unfortunately, if the coin is fair the chances of a head or a tail on the next toss remain the same; both are still 0.5. With a fair coin, the chance of getting a tail is fifty-fifty and remains so no matter what happened on previous throws. Even if we extend the run of heads to 20, 30 or whatever, the probability of getting heads and that of getting tails on the next toss remain equal to each other.

We might, of course, begin to be suspicious of the fairness of the coin if there were an extended run of either heads or tails. In ten tosses of a fair coin the probability that each toss results in a head is

$$\frac{1}{2} \times \frac{1}{2} \times \frac{1}{2} \cdots \times \frac{1}{2} = (\frac{1}{2})^{10} = 0.000977$$

In twenty tosses the corresponding probability is a tiny 0.000000954. Such low values might reasonably make you doubt the fairness of the coin and perhaps the honesty of its owner.

In the opening scene of playwright Tom Stoppard's marvellous piece of theatre, *Rosencrantz and Guildenstern Are Dead*, Rosencrantz has called heads and has been correct in 76 consecutive spins of a coin. Certainly in this case Guildenstern has every right to remark, 'A weaker man might be moved to re-examine his faith, if in nothing else at least in the laws of probability'.

Guildenstern might have done better to ask for the coin to be spun twice and on each pair of spins ask Rosencrantz to call 'head-tail' or 'tail-head' with tail-tail and head-head results being ignored. The game is now fair even if the coin is biased towards coming down heads, as 76 consecutive heads would seem to suggest—even to those undergoing an existential crisis about the laws of probability!

Having Babies

Described by many as 'a wit in the age of wits', John Arbuthnot was born on the 29th April 1667 in Arbuthnot, Kincardineshire, Scotland. A close friend of Jonathan Swift and acquainted with all the literary celebrities of the day, he was also a distinguished doctor and writer of medical works and a physician to Queen Anne. In 1712, he published five statistical pamphlets against the Duke of Marlborough, called *The History of John Bull* (this was the origin of the popular image of John Bull as the typical Englishman). He helped to found the Scriblerus Club and was the chief contributor to the *Memoirs of Martinus Scriblerus* (1741).

Arbuthnot died on the 27th February 1735 in London. Shortly before his death he wrote to Swift, 'A recovery in my case and in my age is impossible;

the kindest wish of my friends is euthanasia.' After his death, Dr. Johnson offered the following epitaph: 'He was the most universal genius, being an excellent physician, a man of deep learning and a man of much humour.'

John Arbuthnot

In addition to his role as clinician and writer, Arbuthnot had a knack for mathematics and an eagerness to apply mathematics to the real world. On the 19th April 1711 he presented, to the Royal Society of London, a paper entitled 'An Argument for Divine Providence, taken from the constant Regularity observ'd in the Births of both Sexes.' In his paper Arbuthnot showed that in each of the 82 years from 1629 to 1710, the annual number of male christenings had been consistently higher than the number of female christenings, but never very much higher. Arbuthnot argued that this remarkable regularity could not be attributed to chance and must therefore be an indication of divine providence. Arbuthnot offered an explanation for the greater supply of males as a wise economy of nature, because males, having to seek out food for their families, are more subject to accidents and diseases. Thus, provident nature, to compensate, brings forth more males. The near equality of the sexes is designed so that every male may have a female of the same country and of suitable age.

Arbuthnot's representation of the chance determination of sex at birth was the toss of a fair coin with one face marked 'boy' and the other 'girl'. Consequently he argued that the probability of observing 82 succesive 'male years' through the operation of chance alone was $(\frac{1}{2})^{82}$, a very small number indeed. Hence he preferred to abandon chance as an explanatory mechanism in favour of divine providence.

In fact, it is now well known that, for whatever reason, more male children are born than females and that the difference varies from country to country and over time. The difference is always relatively small, however. For example, in the United States, during the 20-year period from 1960 to 1979 there were 72,105,000 live births, of which 36,960,000 were male. The relative frequency value for the probability of a boy is therefore

$$\mathrm{Pr(boy)} = \frac{36,960,000}{72,105,000} = 0.51$$

For the purpose of assigning probabilities, however, we will assume that boys and girls are born with equal frequency and that $\mathrm{Pr(boy)}=\mathrm{Pr(girl)}=\frac{1}{2}$. Under this assumption (and ignoring other subtleties like possible biases introduced by the tendency of some families who have had all girls, making another attempt to get a boy), what can we say, for example, about the probability of families of specific sizes having a particular number of boys? To begin let's consider the simplest situation of a family of two children.

Possible make up of family with two children:

Possibility	First child	Second child	Probability
1	Boy	Girl	$\frac{1}{4}$
2	Girl	Boy	$\frac{1}{4}$
3	Girl	Girl	$\frac{1}{4}$
4	Boy	Boy	$\frac{1}{4}$

Thus we have

$$\mathrm{Pr(no\ boys)} = \frac{1}{4}$$
$$\mathrm{Pr(one\ boy)} = \frac{1}{2}$$
$$\mathrm{Pr(two\ boys)} = \frac{1}{4}$$

At the risk of inducing the need for a brief snooze in many readers let's also consider families with three children.

Possible make up of family with three children:

Possibility	First child	Second child	Third child	Probability
1	Boy	Girl	Girl	$\frac{1}{8}$
2	Girl	Boy	Girl	$\frac{1}{8}$

(Continued)

(*Continued*)

Possibility	First child	Second child	Third child	Probability
3	Girl	Girl	Boy	$\frac{1}{8}$
4	Boy	Boy	Girl	$\frac{1}{8}$
5	Boy	Girl	Boy	$\frac{1}{8}$
6	Girl	Boy	Boy	$\frac{1}{8}$
7	Girl	Girl	Girl	$\frac{1}{8}$
8	Boy	Boy	Boy	$\frac{1}{8}$

Thus we have

$$\text{Pr(no boys)} = \tfrac{1}{8}$$
$$\text{Pr(one boy)} = \tfrac{3}{8}$$
$$\text{Pr(two boys)} = \tfrac{3}{8}$$
$$\text{Pr(three boys)} = \tfrac{1}{8}$$

Statisticians have developed a general formula which will give these probabilities in the general case of n children, x of them boys. First, the number of combinations of x boys amongst n children is the binomial coefficient introduced in Chapter 3, namely

$$n!/[x!(n-x)!]$$

For each combination the probability of x boys is $(1/2)^x$ and the probability of $n-x$ girls is $(1/2)^{(n-x)}$. So the probability of the particular combination is given by the product of these two probabilities, that is simply $(1/2)^n$. We now add the probabilities for all possible combinations to get what statisticians know as the *binomial distribution*. This formula gives the probability that in a family with n children, x of them are boys, where x can take values from zero up to n. The formula is

$$\text{Pr}(x \text{ boys in a family of } n \text{ children}) = \frac{n!}{x! \times (n-x)!} \times \left(\frac{1}{2}\right)^n$$

To illustrate the use of this formula, let's consider the probability of finding a single boy in a family of 10 children, a problem particularly dear to the heart of the author. The probability is given by

$$\text{Pr(one boy in family of ten children)} = \frac{10!}{1! \times 9!} \times \left(\frac{1}{2}\right)^{10}$$

Doing the arithmetic shows that the value is approximately 0.01—a one in a hundred chance. (Unfortunately I have always assumed I was one in a million so this comes as a grave disappointment. But perhaps things are not quite as bad as they seem. I was the *last* of the ten children in my family, the first nine being girls, so this at least makes me one in a thousand. And I've ignored the relative rarity of families of ten children—perhaps they're about one in a thousand? In which case, of course, ...)

But my story pales to a mere bagatelle besides the story of the Pitofsky family reported in the *New York Times*. In more than a century, seven generations of the Pitofsky family have produced only sons. In 1959 the forty-seventh consecutive boy in the line was born to Mr. and Mrs Jerome Pitofsky of Scarsdale, New York. Assuming no overlooked girls, this corresponds to one chance in 136 trillion! Maybe, but my advice is to be suspicious, be very, very suspicious.

Tossing coins and having babies seem unlikely bedfellows, but in probabilistic terms they are very similar. The unpredictability of which face will appear on a single spin of a coin mirrors the uncertainty of the sex of a newly conceived child. Both demonstrate that the random variation associated with individual chance events can still lead to long-term regularities, a point to which we shall return throughout the remainder of the book.

Rolling Dice

A throw of the dice will never eliminate chance.
Stéphane Mallarmé

. . . one of my most precious treasures is an exquisite pair of loaded
dice, bearing the date of my graduation from high school.
W.C. Fields

The best throw of the dice is to throw them away.
Advice from an old English proverb

Rolling Dice

Unlike test match cricket, Snakes and Ladders is not a game which reflects
much credit on its inventor, particularly for fathers of small children, since
children seem blissfully unaware of the mind numbing boredom produced
by three or four games in succession. If the result of a test match can partly
hang on the toss of a coin, Snakes and Ladders depends wholly on throws of
a die. St. Cyprian charged Lucifer with inspiring the invention of dice;
perhaps he was right?

As we saw in Chapter 1, gambling with dice was enormously popular well
into the eighteenth century, and for at least 2000 years before. But not until
the sixteenth century does definite evidence suggest that dice players
started to consider in any detail the chances of their getting a particular
result when rolling one or more dice. Perhaps the most famous of the early
investigators was the colourful Girolamo Cardano. Born in Milan on the
24th September 1501, Cardano was the illegitimate son of Facio Cardano
and his mistress Chiara Michina. In fifteenth-century Italy illegitimacy was
no handicap either socially or professionally, and Girolamo Cardano studied medicine in Padua and for a time practised as a country doctor. But
he was better known as an eccentric, a man with a sharp tongue, and a
gambler—facts evident from both his autobiography, *De Vita Propria Liber*

B. Everitt, *Chance Rules*, DOI: 10.1007/978-0-387-77415-2_5,
© Springer Science+Business Media, LLC 2008

(*The Book of my Life*) and his book on games of chance, *Liber de Ludo Aleae*. The former is notable for its frankness:

> I was ever hot-tempered, single minded and given to women. From these cardinal tendencies there proceed truculence of temper, wrangling, obstinacy, rudeness of carriage, anger and an inordinate desire for revenge in respect of any wrong done to me.

In his autobiography Cardano also confesses his immoderate devotion to table games and dice: 'During many years—for more than 40 years at the chessboards and 25 years at dice—I have played not off and on but, as I am ashamed to say, every day.'(Kasparov, Karpov and their contemporaries may object to Cardano equating chess with gambling but in the fifteenth century the game was played fast, and considerable amounts of money changed hands.)

In *Liber de Ludo Aleae* Cardano gives practical advice to the reader, intermingled with preaching morality and recommending prudence, two virtues conspicuous by their absence in his own autobiography. He warns readers that if they are to gamble at all, they should gamble for small stakes and choose opponents of 'suitable station in life'. He also advises on such matters as cheats who use soapy cards and mirrors in their rings to reflect the playing surfaces. But of primary interest for our own story are Cardano's comments about dice, which contain the first expression of the mathematical concept of probability. He clearly states that the six sides of a die are equally likely, if the die is honest, and introduces chance as the ratio of the number of favourable cases to the number of equally possible cases. The following passage illustrates Cardano's understanding:

> One half the total number of faces always represents equality; thus the chances are equal that a given point will turn up in three throws, for the total circuit is completed in six, or again that one of three given points will turn up in one throw, For example I can as easily throw one, three or five as two, four or six. The wagers therefore are laid in accordance with this equality if the die is honest.

Giralamo Cardano, physician, mathematician and eccentric died insane at the age of 75.

As we shall see later, dice *can* be used for serious gambling, but for many people they are associated with less serious games such as Snakes and Ladders and the like (*less serious* is a relative term here as any parent forced to play such games with a 5-year-old will tell you). Most children's board games require the players to throw a six with a die before they can begin. Assuming both a fair die and a fair throwing mechanism (parents will know what I'm getting at here), we might begin by asking, 'How

Giralamo Cardano

many throws, on average, will be needed to get the required six?' To begin to answer this question, we might generate some empirical results by simply tossing a die until a six is obtained and repeating this process a large number of times. Rather than exhaust myself in actually tossing a die many, many times, I simulated 500 sequences of tossing a fair die until a six was obtained on my computer; the average number of tosses required was 5.44, the least, found 98 times, was a single toss; the most, found only once, was 44 tosses. In fact it can be shown that an average of six throws of a die are needed to get a six, but even taking up to about 20 rolls would not make it necessary to ponder on where your opponent had obtained his or her die. (The average number of tosses of a die needed to get a six is another example of a waiting time introduced in the previous chapter.)

Now let's consider some more complicated situations involving two dice. First, we might be interested in examining the sum of the scores of the two faces; the smallest possible score is two, the largest is twelve. But now, unlike tossing a single die, the various possible scores do not all have the same chance of occurring (as was clear to Galileo). That this is so is most clearly seen by tabulating the ways in which each particular score can be achieved (see the following two-dice table).

Die 1/Die 2	1	2	3	4	5	6
1	2	3	4	5	6	7
2	3	4	5	6	7	8
3	4	5	6	7	8	9
4	5	6	7	8	9	10
5	6	7	8	9	10	11
6	7	8	9	10	11	12

Because the number of possible pairings of the faces of two dice is 36 and all possible pairings are equally likely, each outcome has the same probability, namely 1/36. Thus the probability of a particular score is given by the number of ways the score can occur, multiplied by 1/36. For example, the probability of a score of seven with two dice is $6/36 = 1/6$, and the probability of ten is $3/36 = 1/12$. (The same probabilities apply, of course, to the score obtained on rolling one die twice.)

On average how many rolls of two dice will be needed to get two sixes? The result of 500 games of tossing two dice until both showed a six, simulated on my PC, led to an average number of throws of 36.6 with only one roll being needed on 16 occasions and 219 rolls needed once. Mathematics can show that the average number of rolls of two dice needed to get both showing a six is 36. And it would not be *too* unusual to take up to about 100 throws.

(Again, the average number of tosses of two dice until both show a six is an example of a waiting time for some event to happen. If, in general, the event has a probability of, say, p of happening then it can be shown that the average waiting time is just $1/p$. So for one die the average waiting time for a six is six tosses, and for two dice the waiting time for a double six is 36 tosses. There is more about waiting times when rolling dice later in the chapter.)

The number of rolls of a single die needed to get a six and the number of rolls of two dice needed to get two sixes were two aspects of gambling with dice that interested one of the most famous gamblers of the seventeenth century, the Chevalier de Méré. De Méré consistently won by betting even money that a six would come up at least once in four rolls with a single die. He then reasoned that he would also have an advantage when he bet even money that a double six would come up at least once in 24 rolls with two dice since 4 is to 6 (the number of faces on a single die) as 24 is to 36 (the number of pairings of the faces of two dice). Unfortunately, he found that he lost money on this proposition, and wanted to know why. Being a gambler, not a mathematician, he enlisted the help of Blaise Pascal, who was soon able to demonstrate mathematically where Monsieur de Méré's intuition was going wrong.

(1) *Single die*

    (a) Probability of a six = 1/6.

    (b) Probability of other than a six = 5/6.

    (c) Probability of no sixes in four throws,

$$\frac{5}{6} \times \frac{5}{6} \times \frac{5}{6} \times \frac{5}{6} = 0.48$$

    (d) Consequently the probability of at least one six in four throws is

$$1 - 0.4823 = 0.52.$$

(2) *Two dice*

    (a) Probability of two sixes = 1/36.

    (b) Probability of not getting two sixes = 35/36.

    (c) Probability of no throw of two sixes in 24 throws = $\left(\frac{35}{36}\right)^{24}$

    (d) So the probability of at least one throw of two sixes is

$$1 - \left(\frac{35}{36}\right)^{24} = 0.49$$

Pascal's calculations show that the probability of at least one six in four rolls of a single dice is greater than a half (0.52), but the probability of two sixes in 24 rolls of two dice is *less* than a half (0.49). In the first case an even-money bet will, in the long run, lead to a profit; in the second case, however, the gambler will suffer a loss. (If de Méré had been a mathematician, he would have noticed that changing 24 rolls to 25 *would* have made his two-dice-even-money bets successful. It's also interesting to note that he was such an inveterate and attentive gambler that he realized that his even-money bet on requiring 24 throws for at least one pair of sixes was a loser, even though its probability is only a fraction below 0.5.)

Another gambler who decided to consult with a mathematician before making a particular wager was Samuel Pepys. Pepys went straight to the top and, in November 1693, wrote a long, complicated letter to Isaac Newton about a problem involving rolling various numbers of dice. Newton replied some days later pointing out that Pepys's question was not entirely clear but that, as far as he could understand, it came down to asking which of the following three events is most likely:

(1) At least 1 six when 6 dice are rolled

(2) At least 2 sixes when 12 dice are rolled

(3) At least 3 sixes when 18 dice are rolled.

Because 1 is the average number of sixes when 6 dice are rolled, 2 the average number for 12 dice and 3 the corresponding value for 18, the intuitive answer is that the probabilities of the three events must be equal. This was clearly Pepys's assumption, but Newton demonstrated that event 1 has a probability of 0.67, event 2 a probability of 0.62 and event 3 a probability of 0.60. When Pepys found out that the probabilities were not equal he decided to renege on his original bet!

Shooting Craps

The most popular of the casino dice games is *craps*. Played with two dice, craps originated in the Mississippi region of the USA in about 1800, probably as a game played by slaves. The game figures prominently in that wonderful musical *Guys and Dolls*, as when the troubled Nathan Detroit finally manages to arrange his movable crap game in a sewer far below the streets of New York. The scene ends with the legendary Sky Masterson rolling the dice in a thousand-dollar bet against the 'souls' of each of the other gamblers. (This is not quite so Mephistophelian as it seems, because losing entails only their attendance at a meeting of the Salvation Army organized by the lovely Sister Sarah Brown.)

There are a variety of bets that can be made in craps; the most popular is the pass line bet. Here a player wins if his or her score (the sum of the numbers showing on the two dice) is 7 or 11; a player who scores 2 ('snake eyes'), 3 or 12, loses. Any other score is designated a *point*; after getting such a score a player will continue to throw the dice until she or he either scores a seven and loses, or repeats the same point (not necessarily with the same combination of numbers) and wins. The pass line bet is even-money, so that a winning player who staked $10, for example, will win $10. Such a bet is only fair to the player if the probability of winning is 1/2 or greater. Working out the exact probability of winning is not completely straightforward, but for those readers who enjoy this sort of thing, here are the gory details.

We shall use some shorthand here and let p_2 represent the probability of getting a score of two with the two dice, and similarly, p_3, p_4 etc.

(1) The player wins if a seven or eleven is thrown; the probability of this happening is $p_7 + p_{11}$,

(2) The player also wins if a four is thrown and then another four is thrown *before* a seven appears; this can happen if a player

(Continued)

(*Continued*)
throws a four, or throws other than a four or a seven and then gets a four, or gets other than a four or a seven on two rolls and then gets the required four etc., etc., etc. The overall probability is found by summing the probabilities of all these possible results:

$$p_4 + (1 - p_4 - p_7) \times p_4 + (1 - p_4 - p_7) \times (1 - p_4 - p_7) \times p_4 + \cdots$$

(3) The sum of all these terms can be shown to be

$$\frac{p_4^2}{p_4 + p_7}$$

(4) Similarly for a five, six, eight, nine or ten.

Thus the overall probability of winning on the pass line bet is

$$p_7 + p_{11} + \frac{p_4^2}{p_4 + p_7} + \frac{p_5^2}{p_5 + p_7}$$

$$+ \frac{p_6^2}{p_6 + p_7} + \frac{p_8^2}{p_8 + p_7} + \frac{p_9^2}{p_9 + p_7} + \frac{p_{10}^2}{p_{10} + p_7}$$

(5) If we now put in the previously established probabilities for p_7, p_{11} etc., this rather horrible looking expression can be evaluated to give a numerical value of

$$\frac{244}{495} \approx 0.49$$

The probability of winning on the pass line bet is less than 1/2 (0.49), so in the long term the casino will always win—probably not a startling revelation to most hardened crap shooters.

Waiting Times

As with coin tossing, waiting times for different patterns when tossing a die also give some interesting results, although again, how these results are obtained involves some pretty complex mathematics. But, for interest, here are some average waiting times for some patterns when tossing a fair six-sided die;

Pattern 11: average waiting time 42 tosses
Pattern 111: average waiting time 258 tosses

Pattern 1111: average waiting time 1554 tosses
Pattern 1212: average waiting time 1332 tosses
Pattern 1221: average waiting time 1302 tosses
Pattern 123456: average waiting time 46656 tosses

Again the mathematical details behind these results are given in the article by Dasgupta mentioned in Chapter 4.

Dice and their forerunners, astragali, have been used in board games and gambling for thousands of years. Today, however, they are perhaps less popular than some other forms of gambling that we shall examine in the next two chapters. Calculating the relevant probabilities may now become a little more tricky.

Gambling for Fun: Lotteries and Football Pools

6

Introduction

Gambling is the art of exchanging something small and certain for something large and uncertain. Most religious authorities disapprove of gambling and many Islamic nations prohibit gambling altogether. And in most countries gambling is regulated in some way or other. There is no doubt that gambling can, for some people, become addictive and may even lead to tragic social consequences. For the vast majority of people, however, a small flutter on the lottery or football pools does no harm and brings a little fun and excitement into their lives. But even when gambling for fun it may be sensible to be aware of your chances of winning, so read on.

National Lotteries

The first public lottery to pay money prizes was *La Lotto de Fienze*, which began in Florence in 1530; it was soon followed by similar events in Genoa and Venice to raise funds for various public projects. In the United Kingdom the first such lottery took place in 1569

B. Everitt, *Chance Rules*, DOI: 10.1007/978-0-387-77415-2_6,
© Springer Science+Business Media, LLC 2008

principally to raise money for the Cinque ports.[*] There were 400,000 tickets (lots), with prizes in the form of plate, tapestries and money. The ticket sales did not go well, and high-pressure selling methods had to be used; some towns were forced to appropriate public money to buy tickets. The draw took place at the west door of St. Paul's Cathedral. Up until about 1826 the British Government used national lotteries rather than income tax as a means of raising funds.

The current UK lottery, begun in November 1994, consists of selecting six numbers from forty-nine for a one-pound stake. The winning numbers are drawn at random using one of a number of ' balls-in-drum' type of machine, with unlikely names such as Lancelot, Guinevere, and Arthur (the company that runs the lottery is called Camelot). The machine used in a particular draw is itself selected at random by a 'randomly chosen member of the general public'.

The number of combinations (order is not important here) of six numbers from forty-nine is given by the following binomial coefficient (see Chapter 3)

$$\frac{49!}{6! \times 43!} = 13,983,816$$

Hence the probability that a randomly selected set of six numbers will match those chosen by Lancelot, Guinevere or Arthur is 0.0000000715. This is the probability of scooping the jackpot, which is usually around £4–5 million —if there is more than a single winner in any draw, they share the jackpot equally. This tiny probability of winning explains the smug confidence of my statistician friend who insists in writing down six numbers before each draw but never actually places a bet.

It is quite difficult to appreciate just how low a probability of 0.0000000715 really is. It gives rise to what Douglas Hofstadter calls 'number numbness'—the inability to fathom, compare, or appreciate very big or very small numbers. Number numbness may go a long way toward explaining the continuing success and popularity of the lottery. (But the probability of winning the lottery doesn't even begin to compare with the number numbness associated with a probability mentioned in physicist, Murray Gell-Mann's book, *The Quark and the Jaguar*. As an undergraduate, Gell-Mann was asked to calculate the probability that some heavy macroscopic object would, during a certain time interval, jump a foot in

[*] The name of a confederation of maritime towns in southeast Enlgand, formed during the eleventh century to furnish ships and men for the King's service. Dover, for example, is one of the original five towns.

the air as a result of a quantum fluctuation. He found the answer to be around one divided by the number written as one followed by sixty-two zeroes. This really does produce number numbness.)

A comparison of the probability of winning the lottery with some other figures associated with the various types of events that may hasten us to our final rest place is instructive. In an average lifetime of 70 years, the following are the estimated probabilities of dying from various causes:

Cause of death	Probability
Motor vehicle accident	0.017
Drinking chlorinated water	0.00056
Eating 3 oz charcoal broiled steak/day	0.00035
Being killed in a flood	0.000042
Being struck by lightning	0.000035
Being killed by a falling meteorite	0.000000004

Amongst these, the only probability comparable to that of winning in the UK lottery is the probability of being killed by a falling star! It seems that my friend has good reason for his smugness. Astute readers will, of course, quickly see that this comparison—equating *lifetime* risk of death with a probability of winning the lottery in a *single* week with a *single* ticket—is not strictly fair. I make no apologies and invoke author's licence to emphasize a point.

(Incidentally, for those readers who are about to consult their reference books for details of deaths by meteorite, you won't find any. There are no authenticated human fatalities, although automobiles have been struck a few times, most recently by the Peekskill meteorite on 9th October 1992. How then, is the figure in the table arrived at? A certain amount of mathematical wizardry is involved, and a fascinating account is given in an article by Dr. Clark Chapman and Dr. David Morrison published in the January 1994 issue of *Nature*.)

Enthusiastic lottery players will be quick to point out that they are staking only very small amounts in the hope, however remote, of an enormous return. They might also bring it to my attention that they do not need to match all six numbers to win a prize. Tickets matching three or more numbers also win. Let's examine the chances associated with these other possibilities.

(1) Five numbers: the five numbers can be selected from the six winning numbers in six ways, and the remaining selection can be chosen from any of the 43 non-winning numbers. So

(*Continued*)

MAN OF LETTERS

MAN OF LOTTOS

(*Continued*)

there are $6 \times 43 = 258$ ways in which five winning numbers can be selected each with probability 0.0000000715. The overall probability of getting a five-number win is therefore

$$258 \times 0.0000000715 = 0.00000184$$

The typical prize for such a win is £1,500.

(2) Four numbers: here the required probability is

$$15 \times 1118 \times 0.0000000715 = 0.00119$$

The typical prize for a four-number win is £65.

(3) Three numbers: here the probability is

$$20 \times 12341 \times 0.0000000715 = 0.0176$$

For this type of win there is a fixed prize of £10.

Keen gamblers will not have failed to note that the prizes awarded for each type of win fail by a considerable amount to be reasonable returns, given the associated probabilities. Take the three-number win, for example; the chance of winning is about 1 in 50, but the return on a one-pound stake is only £9, the equivalent of 'odds against of only 9 to 1. On average, the UK National Lottery returns only 45% of money staked.

Despite the completely random nature of the draw, many participants in lotteries adopt some particular 'strategy' when making their selections. The majority of sets of six numbers chosen are certainly not random, because many people use birthdays and other special dates, as the basis of their choices. It has also been reported that between 10,000 and 30,000 players per week select the numbers $\{1, 2, 3, 4, 5, 6\}$. And Camelot have stated that 45% of players use the same numbers each week and that workplace or family lottery syndicates tend to stick with 'their' numbers. All of this suggests that a rational strategy is to try to win the jackpot in those weeks when it will not be shared among several winners by choosing six numbers genuinely at random, but reject any combination (such as 1 to 6 or all six numbers under 31) that there is good reason to believe will be chosen by many others. With this in mind it's almost certainly an advantage to have a 'lucky dip' selection—that is, let the lottery machine choose your numbers. *No* strategy, of course, alters your chance of hitting the jackpot. This remains almost zero! In fact Dr. John Haigh of the University of Sussex in the UK has calculated that you shouldn't buy a lottery ticket until 7.10 p.m. if you want to have a better chance of winning than of dying before sales close at 7.30 p.m.

Just how small the chance is of scooping the jackpot in the UK lottery can also be seen if we consider how long, on average, a player buying a single ticket a week would have to wait before collecting his or her very large cheque. For this, cast your mind back to Chapter 5 and the discussion there of tossing a single die and two dice. When tossing a single die the average number of tosses before the appearance of a six was found to be six and when rolling two dice the average number of tosses needed to get a double six was given as 36. And, in general, the average waiting time to an event with probability p is $1/p$. In the UK lottery the probability of matching all six numbers and winning the jackpot is about one divided by 14 million, so the average waiting time for a jackpot win for the single ticket a week player is simply one over this figure, giving the average waiting time to winning the jackpot of 14 million weeks, about 270,000 years! Even if you were as rich as say Richard Branson and invested say £1000 weekly on the lottery, the average time before you might become a jackpot winner would be 270 years.

But, of course, some people do win the jackpot and change their lives. One of the strangest stories involves a pub syndicate in Kent who, in September 1997, won nearly £11 million. The syndicate bet on all combinations of six numbers from eight they selected, at a cost of £28. One of the syndicate was entrusted to fill in the 28 different combinations on the lottery tickets. But he made a mistake, probably the most fortunate mistake he ever made; he entered only 27 different combinations, duplicating one for the 28th entry.

The duplicated ticket was the jackpot-winning combination so they won two jackpot prizes, a total of £10.8 million with the remainder of their win made up of a number of smaller prizes for five- and four-number matches.

Football Pools

Although the National Lottery is now the number one gamble in the UK, many punters still retain a degree of affection for the old football pools, particularly the 'treble chance' bet in which the participant tries to select, from about fifty Saturday soccer matches, eight that will result in draws—I shall ignore the recent innovation of dividing these into no score draws and score draws. Football pools are so called because the entry fee money is all *pooled* and the prize money is a share of that total amount. (British football pools have a large international participation which makes this account of them a little less parochial that you might have thought.)

In the UK, the biggest and most successful of the football pool operators is Littlewoods Pools now a multi-million-pound enterprise although declining a little since the introduction of the lottery. The foundations of Littlewoods Pools were humble. The company was launched in 1923 as a part-time venture by three young Manchester telegraphists with a total capital of just £150. Because they thought that their idea might be frowned upon by their employers, they decided not to use any of their own names. But one of the partners, Harry Askham, had adopted the surname of an aunt who raised him—he had been born Harry Littlewood. Four thousand copies of the first coupon were printed and distributed outside the Manchester United football ground. Only thirty-five of them were returned, and the total amount of money invested was just over £4. The first payout was just over £2. A little later 10,000 coupons were distributed at a major football fixture in Hull—only one was returned. At the end of the first year, prospects for the venture looked bleak and two of the three partners pulled out. The one who remained, John Moores, went on to become a millionaire, and Littlewoods Pools became for many years an established feature of the British way of life. Up until the 1990s almost the entire population of the UK 'did the pools' and pools results were published in most national newspapers a day or two after the Saturday on which matches were played. But earlier than this, on the Saturday afternoon at about 5 p.m. when the matches were finished, the results were broadcast on the radio and we all sat checking our football coupons hoping for a big win.

The football pools did not fall under gambling legislation because they claimed to be competitions of skill, rather than chance. In the 1950s the top dividend was £75,000 but recently it has climbed well over the million-pound mark. The first football pools millionaire was Nursing Sister Margaret Francis in 1984. One winner who gained a degree of notoriety was Viv Nicholson who after winning £152,319 in 1961 declared that she was going to spend, spend, spend. And this she did with some success, eventually being made bankrupt.

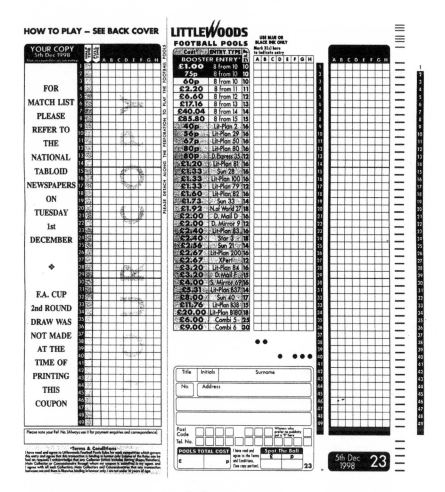

Football pools coupon

Although gambling on the football pools might be considered to involve an element of skill singularly absent from playing the lottery, it is unfortunately not possible to forecast draws with any accuracy. Consequently, many who take part simply pick eight numbered matches at random.

Evaluating the chances of being a winner on the pools is slightly more tricky than for the lottery, because that chance largely depends on how many of the fifty matches actually result in draws. If, for example, all fifty matches resulted in draws on one miraculous Saturday, every gambler would 'win' but win nothing. To begin, let's consider a week in which only eight draws occur. Now there is only a single way in which a winning set of matches can be selected, and the number of combinations (again order is not important) of eight numbers from the fifty on the pools coupon is

$$\frac{50!}{8! \times 42!} = 536,878,650$$

The corresponding probability of becoming an overnight millionaire is 0.000000001862, about 50% of the chance of being killed by a tumbling meteorite.

But what about weeks when there are more than eight drawn matches? Suppose for example, there are twelve draws in a particular week. The number of ways matches can be selected from the twelve draws is

$$\frac{12!}{8! \times 4!} = 495$$

Each of these ways has an associated probability of 0.000000001862, so the total probability of winning in such a week is simply $495 \times 0.000000001862 = 0.00000092$, about thirty times less than being struck dead by a lightning bolt. And winners in such a week will, of course, receive lower prizes than in a week with only eight draws, in which there is very likely to be only a single winner. These small probabilities will come as no surprise to readers, who, like the author, have diligently posted off their coupons for several decades with no return. But, at least forecasting draws on the football pools does involve something that may be of interest to many participants: the results of football matches. Such redeeming interest is woefully wanting in the lottery, which, instead, provides a completely mindless pursuit for the gullible who seem unaware that they are almost as likely to be the victims of a cosmic bolt from the blue as to become instantly rich. (In case this sounds like pompous moralizing, I should confess that 'Get lottery ticket' still appears as an item on my supermarket list every Saturday morning. After all, it's for a good cause—making money for Camelot!)

Both lotteries and football pools nicely illustrate 'the triumph of hope over experience'. The chance of winning the jackpot is tiny for participants in either activity. But most players probably welcome the (generally short-lived) excitement of checking their lottery ticket or football coupon on a Saturday evening, and for the vast majority, the gamble is just a bit of fun. 'Serious' gamblers worth their salt would not, of course, be caught indulging in such trivial pursuits. Their belief is that there are other, more rewarding, activities in which to 'invest'. In the next chapter, well try to demonstrate how wrong (and occasionally how right), they can be.

Serious Gambling: Roulette, Cards and Horse Racing

A man's idee in a card game is war—crool, devatatin' and pitiless. A lady's idee iv it is a combynation iv larceny, embezzlement an' burglary.
Peter Finley Dunne

The way his horses ran could be summed up in a word. Last. He once has a horse who finished ahead of the winner of the 1942 Kentucky Derby. Unfortunately, the horse started running in the 1941 Kentucky Derby.
Groucho Marx

Yes, I wish I had played the black instead of the red at Cannes and Monte Carlo
Winston Churchill in reply to the question ' If you had it all to do over, would you change anything?'

Introduction

Serious gambling is big business. In the United States, for example, gambling is at least a $40-billion-a-year industry. Casino gambling is the most popular leisure activity—in 1996, nearly thirty million people visited that Mecca of gambling, Las Vegas. The visitors' average gambling budget for the trip was about $600 and, again on average they spent 4 hours a day gambling. Somewhat sadly, this huge industry is built on the losses of occasional gamblers who know little or nothing about the hard mathematical and psychological facts of the games on which they now and then wager some money.

Serious gambling might, of course, involve some trivial event. Betting on which of two raindrops on the outside of a bus window gets to the bottom first would be a serious enough prospect if the stakes were £10,000. But here, attention will be restricted to horse racing, cards and roulette, because each can be used to illustrate further aspects of the operation of chance,

B. Everitt, *Chance Rules*, DOI: 10.1007/978-0-387-77415-2_7,
© Springer Science+Business Media, LLC 2008

although betting on horse racing or cards involves some level of skill noticeably absent from betting on the spin of a roulette wheel.

Roulette

Roulette is played at a table with a wheel and a betting area. The wheel rotates around a vertical axis and is located in a shallow bowl with a wall curved towards the inside. The wheel and the bowl are so designed that a small ball can be spun on the inside of the wall, without flying outside, and such that after several rotations the ball finally drops into one of the pockets. The pockets are numbered and are painted so that red and black alternate. Gambling historians disagree over the origins of the game. According to some, Blaise Pascal invented roulette in 1655 during his monastic retreat. Others claim that it was invented by a French monk to help relieve the monotony of monastry life. A third suggestion is that roulette originated in an old Chinese game the object of which was to arrange thirty-seven statuettes of animals into a magic square. According to this account, the statuettes were eventually transformed into the numbers 0 to 36 and arranged haphazardly along the rim of a revolving wheel by French Dominican monks. (The only consistent theme in the three suggestions is the monastic setting.)

Roulette wheels differ in their construction. In most North American casinos they have two 'zeros', 0 and 00. In Europe and South America, however, most wheels have only one zero. Because many bets are automatically lost if the result of a spin is 0 or 00, the game is substantially more favourable for the player when the wheel has only a single zero. Roulette bets are made by placing chips (or occasionally currency) in certain parts of the betting area. Essentially, bets differ only in the number of numbers which are covered in the bet. The simplest (though perhaps not the wisest) bet, for example, is on a single number—say 32. If the ball lands on 32 the player wins, and the bank pays odds of 35 to 1. This corresponds to a probability of 1/36. Unfortunately for the gambler, there are either 37 or 38 numbers on the wheel (depending on where in the world he is playing) so the correct odds for a fair game would be either 36 to 1 or 37 to 1. The presence of the 0 gives the bank a small but distinct advantage; the presence of both 0 and 00 confers an even greater advantage on the casino. We can assign numerical values to this advantage very simply. Imagine, for example, a roulette wheel in a North American casino; the wheel has a 0 and 00. When you place a one-dollar bet on a single number, the probabilities of winning and losing are

$$\text{Probability of a win} = \frac{1}{38}$$

$$\text{Probability of a loss} = \frac{37}{38}$$

If the ball lands in your chosen number you win \$35, so the expected return on one dollar is

$$35 \times \frac{1}{38} - 1 \times \frac{37}{38} = -\frac{2}{38} = -0.0526$$

So the expected return on \$100 is about −5.26 dollars, where the negative sign, of course, indicates a loss to the player. The casino's advantage is about 5.3%. With a single zero, the bank has an edge of about 2.7%. The expected return would be zero (indicating a fair game) only if the odds for a single number on a wheel with 0 and 00 were 37 to 1, and on a wheel with only 0 were 36 to 1, rather than the 35-to-1 odds actually given. (A player could, of course, put his or her single number bet on 0 or 00 if available and would then win if the ball fell in the chosen value—the same negative expectation of gain, however, would still apply.) Clearly, casino operators have learnt that it is dangerous to give players an even break, because in the long run they would break even! 'Never give a sucker an even break' as somebody once so eloquently remarked.

Over the years, numerous 'systems' have been developed for making roulette a successful enterprise for the player rather than the casino. In the Biarritz System, for example, a player must, before placing any stakes, note the results of at least 111 spins of the wheel. Then, having noted what numbers come up with less than a certain frequency, the player bets on those numbers, the implicit assumption being that their former rarity will be offset by an immediate glut. The reasoning is similar to that dealt with in Chapter 4 when discussing long runs of heads or tails when tossing a coin. And once again, it is a manifestation of the Gambler's Fallacy. If such a system *was* successful it would simply mean that the casino should consider buying its roulette wheels from another manufacturer. All similar systems are flawed for the same reason. As John Scarne, America's self-styled doyen of gambling, puts it 'No roulette system is worth the price of yesterday's newspaper'. Furthermore, the bank's favourable advantage of 0 and (perhaps) 00 will, in the long run, break all players. The best way to avoid losing at roulette is to stop playing the game.

No account of roulette, however brief, would be complete without some mention of the casino at Monte Carlo. So here's a remarkable fact. The longest series recorded in Monte Carlo is red coming up twenty-eight times in succession. Assuming that the probability of a red number is $\frac{18}{37}$, (remember that zero), the probability of this sequence occurring is

$$\left(\frac{18}{37}\right)^{28} = 0.00000000173$$

or about 578 million to 1! Anybody who placed a small bet at the beginning of this sequence and doubled up for the whole run would never have had to work again. What actually happened, however was that after ten or so reds in succession, players rushed to bet on the black, doubling and even tripling their stakes. In the end, as usual, the only winner in the end was the casino.

Such a long sequence of one colour is impressive, but if you are tempted to invoke the hand of God, the intervention of aliens, or some other such fanciful explanation, consider the following;

- Monte Carlo has several tables in daily use.
- Each wheel makes an average of 500 spins per day.
- The wheels are in action almost every day of the year.

A simple calculation shows that if you wait around for about 100 years, the probability of observing a run of twenty-eight reds becomes very high. One more fact: the casino at Monte Carlo has been in operation for almost 100 years.

Playing Cards

When I was a child one of the most exciting treats at Christmas was being allowed to play cards with the grownups in the evening. The games we played, such as Sevens and Newmarket, were not themselves particularly exciting but we played for money. For a 10-year-old this was living on the edge. I loved the thrill of putting my carefully accumulated pocket money at risk for the possibility of becoming rich—though in fact we played for pennies, and the greatest gain or loss over the course of an evening was rarely more than a pound. On some occasions, of course, I lost, and I have never been allowed to forget the fuss I made until some kindly relative reimbursed me.

In my teenage years I continued to play cards at school, but the games now involved, such as three-card brag, could mean the rapid loss of hard-earned wages from paper rounds, and no avuncular figure stood ready with a broad shoulder to cry on and a thick wallet to insure against bankruptcy. Later on in life, I progressed to playing frequent games of poker, naively thinking that my training as a statistician would give me an advantage over more regular players unfamiliar with the mathematical aspects of probability. I was wrong! Fortunately, I learnt quickly that theoretical knowledge is no substitute for years of practical experience gained in ill-lit, smoke-filled rooms, and I turned to other pursuits which offered to increase rather than decrease my bank balance.

"When you've finished playing
with the children"

Playing cards is an extremely popular recreation, and not primarily for hardened gamblers. A survey carried out some years ago found that 83% of families surveyed regularly played cards and that cards were found in 87% of homes. Some fifty million decks of playing-cards are purchased in the United States each year. Having probably originated in China, playing-cards are believed to have arrived in Europe from the East, specifically as derivatives of the cards used by the Mamelukes of Egypt. An almost complete pack of Mameluke playing-cards was discovered in the Topkapi Sarayi Museum, Instanbul by L.A. Mayer in 1939. The pack dates from somewhere around the twelfth century and consists of fifty-two cards with suits of swords, polo-sticks, cups and coins. Numerals from 1 to 10 are used, and court cards are labelled *malik* ('King'), *naib malik* ('Deputy King') and *thani naib* ('Second Deputy'). It is likely that such playing-cards arrived first in Italy and that their use spread rapidly. Cards are first mentioned in Spain in 1371, are described in detail in Switzerland in 1377, and by 1380 are reliably reported from places as far apart as Florence, Basle, Regensburg, Brabant, Paris and Barcelona. Early cards were individually handmade and painted, which made them expensive to produce and may at first have restricted the market to the well-to-do. But by the fifteenth century, cheaper versions were available and cards were popular at all levels of (at least) urban society.

There are many card games but here we shall restrict our attention to just three: poker, bridge and blackjack. Poker and blackjack are generally thought to be the choice of most serious gamblers with bridge often thought to be the choice for those looking primarily for an intellectual challenge. In fact, many devotees of both poker and bridge maintain that poker demands considerably more skill than bridge. And there are some that would have us believe that blackjack may, if played properly, not be gambling at all.

For none of the three games is it essential to have a technical knowledge of probability theory. As in all good games, there are many attributes that make a good poker, bridge or blackjack player. Understanding of the influence of chance is just one, and most competent players learn enough largely through a combination of rote and experience. Nevertheless, a somewhat more detailed grasp of the way the probabilities of particular hands can be calculated might benefit many players. This is the modest aim of the next three sections. In no way are these sections intended to be guides to playing any of the three games.

Poker

Poker is probably the most popular card game in the world, in terms of both the amount of money that changes hands at the game every year and the number of players. According to John Scarne, poker seems to have been developed by card sharps working the Mississippi river boats in the early nineteenth century. Although most dictionaries and game-historians suggest that the name comes from the early-eighteenth-century French game, *Poque*, Scarne believes it arises from underworld slang and comes from the pickpocket's term for pocketbook or wallet—*poke*.

The most common form of poker involves players being dealt a hand of five cards, some of which they may later exchange in an attempt to improve their hand. Hands are divided into groups such as no pair, pair, three-of-a-kind, straight, flush, full house, four-of-a-kind, straight flush and royal flush. These groups are then ranked in relative value according to their frequency of occurrence: the hands that can be expected to appear most often have the lowest rank; those that appear least often the highest. Thus, for example, a final hand of three-of-a-kind will beat a final hand of a pair. A good poker player needs to have a fair idea of the probabilities of various poker hands, but, as I found to my cost as a young man, poker is primarily a game of strategy, deception and psychology. On some occasions, knowing the exact strength of your hand or the exact chances of bettering your hand by drawing a particular number of cards will not help you too much, because the playing habits of your opponents may destroy the best-laid mathematical plans. A big raise from a calm, controlled player gives quite a different message from the same raise made by a drunk who has already tried to bluff the last half-dozen hands.

But having some knowledge of how to calculate the probabilities involved is useful. Let's begin by considering how many possible 5-card poker hands can be dealt from a 52-card deck. The answer is found by using the combination formula we met in Chapter 3. The number is given by

$$\text{Number of possible poker hands} = \frac{52!}{5! \times 47!} = 2,598,960$$

Now consider the probability of being dealt particular types of hands, such as four-of-a-kind (four cards of equal face value), full house (one pair and one triple of cards with equal face values), and so on.

(1) *Four-of-a-Kind*

   (a) Consider first a hand consisting of four aces. The aces can only be selected in one way (there are, after all, only four aces in the pack); the remaining card can be selected from the 48 'non-aces' in 48 ways. Consequently there are 48 such hands.
   (b) The same reasoning applies to each other four-of-a-kind hand, so in total there are $13 \times 48 = 624$ of this type of hand.
   (c) The required probability then is simply

$$\Pr(\text{Four} - \text{of} - \text{a} - \text{Kind}) = \frac{624}{2598960} = 0.00024$$

(2) *Full House*

   (a) First, there are thirteen possible denominations of cards for the pair. From the four cards of a particular denomination, the pair can be selected in $\frac{4!}{2! \times 2!} = 6$ ways. The denomination of the required triple can now be chosen in 12 ways, and for a particular denomination the triple can be selected in $\frac{4!}{3! \times 1!} = 4$ ways. The total number of such hands is therefore

$$\text{Number of full house hands} = 13 \times 6 \times 12 \times 4 = 3,744$$

   (b) So in this case the required probability is

$$\Pr(\text{Full House}) = \frac{3744}{2598960} = 0.0014$$

(Continued)

(*Continued*)

(3) *Pair*

    (a) Now lets find the number of hands containing a single pair. This is a hand with the pattern AABCD, where A, B, C and D are from distinct denominations of cards. Consider first a pair of aces. The two aces can be chosen from the four aces in the pack in $\frac{4!}{2! \times 2!} = 6$ ways. The remaining three cards must be chosen from the 12 remaining denominations, and a single card drawn from the four cards of each of the three selected denominations. The number of ways these three cards can be drawn is therefore

$$\frac{12!}{3! \times 9!} \times 4 \times 4 \times 4 = 14,080$$

    (b) Now the number of hands containing a pair of aces is simply $6 \times 14,080 = 84,480$ and so the number of hands containing a single pair of any type is $13 \times 84,480 = 1,098,240$.

The probability of a pair is consequently 0.423.

Keen readers can, in a similar fashion, calculate the probability of two pairs as 0.048 and of a triple as 0.021. The majority of readers can take my word that these are the correct values. A summary of the probabilities of various hands in 5-card poker follows and their frequencies are given in the accompanying figure.

Hand	Description	Probability
Single pair	AABCD	0.423
Two pair	AABBC	0.048
Triple	AAABC	0.021
Full house	AAABB	0.0014
Four of a kind	AAAAB	0.00024
Straight	Five cards of any suits in sequence	0.0039
Flush	Five cards from the same suit	0.0020
Straight flush	Five cards in sequence from the same suit	0.000015
Royal flush	Ten, jack, queen, king and ace of one suit	0.000002
None of the above		0.501

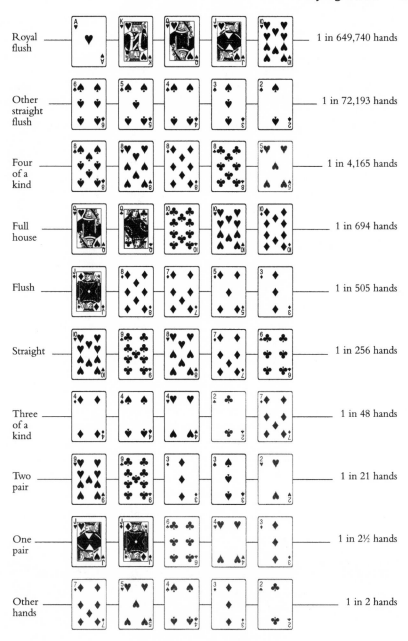

Frequency of poker hands

(Note to lottery players in the UK: The chance of getting a royal flush at poker is nearly 30 times the chance of winning the National Lottery with a single ticket.)

Perhaps more important than the probabilities associated with the original hands dealt are the probabilities of improving a hand when drawing cards. Consider, for example, a player whose original hand contains a pair; she might be interested in the chances of making four-of-a-kind if she draws three cards to the pair and in knowing how that strategy compares with the alternative of drawing only *two cards*, keeping her pair and one of the other cards. Without going into details, it turns out that the chance of improving the hand containing a pair to four-of-a-kind is almost three times higher when three cards are drawn. But good poker playing demands that players occasionally use the probabilistically less advantageous strategy in order to keep opponents in doubt about their playing habits.

Contract Bridge

The essential features of the game are as follows:

- Four persons play, two against two as partners—the partnerships are designated *North-South* and *East-West.*
- A full 52-card pack is dealt out one at a time clockwise around the table so that each player holds thirteen cards.
- The object of the game is to win tricks, each trick consisting of four cards, one card played by each player.
- One suit may be designated as the trump suit; that is, any card in that suit will take any card of the other suits.
- All four players bid for the right to name the trump suit (or of determining that a hand will be played with no trump suit). One member of the higher-bidding team plays the hand with his/her partner's hand (dummy) exposed and attempts to fulfil the 'contract' (win the number of tricks) to which the team committed itself during the bidding.
- A scoring system has implications for the bidding that takes place.

Bridge was probably born of three-hand whist games. Inveterate whist players, unwilling to forego their game merely because there were only three available players, played a game called dummy (with one hand exposed) long before any bridge game was known. Bridge itself probably originated in the late nineteenth century in Turkey and Greece. Though bridge was initially thought to be less scientific than whist (wherein all four hands were hidden), it soon became apparent that exposure of dummy greatly increased the possibilities of skillful play, and by the dawn of the twentieth century nearly all the leading whist players had succumbed to the attractions of the new game. Bridge is a challenging game with many aspects and a huge literature. Here we simply derive some probabilities associated with particular hands.

We might begin by asking about the number of possible bridge hands. The answer is a staggering $53, 644, 737, 765, 488, 792, 839, 237, 440, 000$. The number of possibilities involved in the current UK lottery pales by comparison. How is the enormous number arrived at?

(1) The number of ways in which the first player's 13 cards can be allocated is:

$$\frac{52!}{39! \times 13!} = 635,013,559,600 \ (A)$$

(2) The number of ways the second player's hand can be chosen is:

$$\frac{39!}{26! \times 13!} = 8,122,425,444 \ (B)$$

(3) The number of ways for the third player's cards to be selected is:

$$\frac{26!}{13! \times 13!} = 10,400,600 \ (C)$$

(4) This leaves the remaining 13 cards for the fourth player.
(5) The total number of hands is therefore simply

$$A \times B \times C \times 1$$

which gives the truly gigantic number above.

Let's now look at the probability associated with all four players being dealt a hand with thirteen cards in one suit. There are twenty-four ways in which all four players can receive a complete suit ($4 \times 3 \times 2 \times 1$). Hence the probability of such a distribution of cards (a 'perfect deal') is really very small—approximately twenty-seven zeros followed by a 5 Such a probability makes winning the lottery look almost like a sure thing. It is approximately the same as the probability of a run of 91 heads when tossing a fair coin. Richard Dawkins, in his best-selling book *The Blind Watchmaker*, even suggests a special name for such a tiny probability, the *dealion* (the rest of the book is more inventive). Somebody with time on his hands once calculated that if the entire adult population of the world played bridge continuously for 10 million years, the odds would still be 10 million to 1 against such a set of hands. And yet, despite such enormous odds against, several people have claimed to have witnessed a perfect deal. They include

- Dr. R.S. Le May, of Tunbridge Wells, England in 1952.
- Mr. A.J. Smith, a bridge writer, of Welwyn Garden City, England, recalled in 1952 that 'in the fleeting span of my own lifetime I have encountered 5 or 6 instances [of perfect deals].'
- Mr. R.R. Thomas, of London, England, said in 1952 of his perfect deal that 'the pack was not a new one and was properly shuffled, and cut and dealt.'
- Mrs. Irene Motta, of Cranston, Rhode Island, 1954.
- Mrs. E.L. Scott, of Frankfort, Kentucky, in 1958.
- Mrs. Gordon Hay, of Kanakee, Illinois, 1963.

Given the odds against a perfect deal, this probably incomplete collection of six claims of its occurrence in the space of just over a decade beggars belief. Possible polite explanations include poor shuffling and new decks, but it is more likely that the claimants have been duped by pranksters. When an event with a probability of a dealion happens, fraud is a far more likely than a miracle.

Let's move on to some other more realistic and helpful calculations. The distribution of aces is important in most bridge games, so we might ask, 'What is the probability that the bridge hands of North and South together contain all four aces?

(1) The trick here is to consider the *combined* North and South hands. We then first calculate how many hands are there in total—the answer is

$$\frac{52!}{26! \times 26!}$$

(2) Now the number of hands containing four aces is found by multiplying the number of ways the four aces can be selected from the four in the pack (one way) and the number of ways the remaining 22 cards of the combined hand can be chosen from the 48 non-aces, $\frac{8!}{22! \times 26!}$ ways. This gives the required probability as

$$\frac{\frac{48!}{22! \times 26!}}{\frac{52!}{26! \times 26!}} = 0.06$$

(3) Just over a 1 in 20 chance.
(4) Similarly we can find the probabilities of 3, 2, 1 and zero aces as 0.25, 0.39, 0.25 and 0.06.

A Yarborough is a hand that contains no card higher than a nine. (The name for this rather poor hand is said to go back to a Lord Yarborough, a whist player who was prepared to bet 1000 to 1 against any given hand being so bad that it contained no card above a nine.) What is the probability of being dealt such a hand? Because there are 32 cards in the pack between two and nine, the probability is simply

$$\frac{\frac{32!}{13! \times 19!}}{\frac{52!}{39! \times 13!}} = 0.000547$$

Lord Yarborough's offer of 1000 to 1 against such a hand is hardly generous—he should have offered almost 2000 to 1.

Blackjack

Blackjack or '21' is one of the most popular casino games. The French are usually credited with inventing blackjack which is believed to be an off-shoot of French card games 'chemin de fer' and 'French Ferme'. The game 'vingt-et-un' (twenty-and-one) originated in French casinos around 1700 A.D. and migrated to the USA in the 1800s. When blackjack was first introduced into the United States it was not particularly popular, so casinos and gambling joints offered various bonus payouts to get players to the tables. One such bonus was a 10 to 1 payout if a player's hand consisted of the ace of spades and a black jack; this hand was called a blackjack and the name has stuck even though the bonus payout' was soon abolished (I wonder why?).

In blackjack each player is dealt two cards and is then offered the opportunity to take more. Cards two through ten are worth their face value, and face cards (jack, queen, king) are also worth ten. An ace's value is 11 unless this would cause the player to exceed 21, in which case it is worth one. The hand with the highest total wins as long as it does not exceed 21; a hand with a higher total than 21 is said to bust. When playing in a casino each player makes an initial bet and then the dealer gives two cards to each player including himself. One of the dealer's two cards is face up so all the players can see it, and the other is face down. At this stage each player has a variety of options including taking no more cards, and taking another card. It can be shown that the house advantage, gained by the dealer being the last player to act, is about 8%; by acting last, all other players will have already made their decisions and might have already best before the dealer has his or her turn.

In the simplest situation where there is only a dealer and one player and the dealer is using only one deck, it is straightforward to evaluate some probabilities that may be helpful. For example, suppose the player currently holds a queen, a two, a four and an ace, and the face up card of the dealer is a four. The best estimate of the probability of the player achieving 21 points (receiving a four) is simply the likely number of fours still in the pack divided by the total number of cards left in the pack, i.e.

$$\text{Probability of receiving a four} = (4 - 2)/(52 - 5) = 0.043.$$

Similarly, the probability of achieving 20 points (receiving a three) is $4/47 = 0.085$, the probability of receiving a two and scoring 19 points is $3/47 = 0.063$ and the probability of being dealt an ace and scoring 18 points is $3/47 = 0.063$. So a reasonable guess at the probability of achieving 18, 19, 20 or 21 points is $0.043 + 0.085 + 0.063 + 0.063 = 0.25$. Since any other total than 18, 19, 20 and 21 will mean that the player has bust, the probability of busting is about 0.75.

Blackjacks popularity as a casino game reflects perhaps that skilled players can lower the odds in favour of the casino and increase their own odds of winning. And by a technique known as card counting it may even be possible to gain a small advantage over the bank. Card counting was originally suggested in the 1960s by Edward Thorp a mathematics lecturer then at MIT and was put to the test by Thorp and Emmanual Kimmel, a New York business man in Reno, Nevada. Kimmel bankrolled Thorp to the tune of $10,000 and eventually they more than doubled their money.

Card counting is described in detail in Thorp's book, *Beat the Dealer*, but in essence it involves the player keeping track of the cards that have already been played and so knowing when the cards in the remaining deck are to the player's advantage. The fundamental principle behind counting cards in blackjack is that a deck of cards with a higher proportion of high cards (tens and aces) to low cards is good for the player, while the reverse is true for the dealer. A deck rich in tens and aces improves the player's odds because blackjacks become more common. There are several card counting systems which do not require the player to remember which cards have been played. Rather, a point system is established for the cards and the player keeps track of a simple point count as the cards are played out from the dealer.

Since card counting is not illegal, so how do casinos avoid loosing money? Some casino owners prevent card counters from playing by simply not allowing likely suspects into their premises. In other casinos, suspected card counters are subject to minor harassment to break their concentration and, in some others, cards are shuffled more often or when a player increases his stake. But, even as recently as the 1990s, students from MIT took on the Las Vegas casinos at blackjack and came home with millions of dollars. It seems likely, however, that casinos will, in the end, arrive at a system that will stop their losses to card counters and that Edward Thorp's scheme for making money playing blackjack will sadly become only a memory.

Horse Racing

Many years ago, a friend of mine worked as a psychologist in the University of one of the United Kingdom's major cities. Sid (not his real name) loved to back horses. A visit to the betting shop gave him as much pleasure as the rest of us get from eating out at a Michelin three-star restaurant, say, or being present on those increasingly rare occasions when England trounces Australia at cricket. But, along with loving a bet, Sid also took great interest in the horses themselves—he was particularly fond of the great Nijinsky. Bought as a yearling in Canada for $84,000 by Charles Engelhard, Nijinsky proved to be one of the great racehorses of all time. Between 1969 and 1970, he won eleven consecutive races including the Triple Crown of the Epsom Derby—the King George VI and Queen Elizabeth Stakes at Ascot and the St. Leger at Doncaster. Three weeks after victory in the St. Leger, Nijinsky was entered for the Prix de L'arc de Triomphe at Longchamp. The horse was odds-on favourite, of course, and so for a chance at a decent return one had to make a large bet. Sid decided that this was the time for decisive action, and he tried to persuade his university accountant to allow him to withdraw all of his pension contributions so that he could bet on Nijinsky. But, despite all Sid's arguments and a detailed account of the great horse's history, the accountant was unmoved, and no pension contributions were forthcoming. Sid was furious—at least until the race was over. The following is taken from Lester Piggott's autobiography; Lester was Nijinsky's jockey on the fateful day.

> The paddock at Longchamp is an amphitheatre, with terraces commanding a view for thousands of people. By October 1970 Nijinsky was the most famous racehorse in the world, and it seemed as if racegoers had come from around the globe to see him. There were only fourteen runners in the Arc, but in addition to their connections the paddock was swarming with camera crews from several different countries, who were causing havoc. A Japanese film crew homed in on poor Nijinsky, and one of them even thrust a microphone under the horse's mouth. Even the best efforts of Vincent's [*Vincent O'Brian*, the horse's trainer] travelling head man, Gerry Gallagher, could not deter these people, and by the time I entered the ring with the other jockeys Nijinsky—always a temperamental horse—had flipped. He was awash with sweat, snorting and dancing around, a look of terror in his eyes: I had a job to get up into the saddle. In the parade and going to the start I was increasingly unhappy with his state of mind, and felt that he must have used up a great amount of nervous energy.

An additional problem was that I was drawn on the outside, and in France you have to keep your position for over a furlong before tacking over towards the inside rail. Once the race was under way, Nijinsky found it difficult to keep his pace early on and stay up handy as he normally could. Half a mile from home he was still towards the rear of the field, but made a forward move when I asked him approaching the short final straight—where his finishing run was blocked by four horses in line ahead of us. I had to pull to the outside, thus losing a little ground, and by the time I straightened him up to go for the post Yves Saint-Martin and Sassafras were still in front on the rails. Nijinsky put his all into pulling back Sassafras and with a hundred yards to go just got his nose in front, but then he faltered and began to hang to the left. He could find no more, and we were beaten a head.

So thanks to a sensible accountant. (Is there any other kind?) Sid's pension fund survived. Unfortunately, upon his wife hearing of his efforts to stake everything on Nijinsky, his marriage did not.

Lester Piggot on two-year old Nijinsky, winning the Dewhurst Stakes at Newmarket in 1969

Horse racing in some form is at least 6000 years old. A carved shell cylinder which dates from 4000 B.C. or earlier shows Assyro-Babylonian warriors using horse-drawn chariots in battle and it seems safe to assume that these soldiers also staged chariot races. The earliest full-length account of a chariot race may be found in Book XXIII of the *Illiad*. First prize included 'a woman skilled in women's work'. Post positions were determined by lot, a patrol judge was sent out, and Antilochus received instructions from Nestor before the race. (He was subsequently disqualified from second place because of having followed them too well.)

The first formal mounted race on record was one that took place in the thirty-third Olympiad in Greece, in about 624 B.C. In Rome each of the various racing factions had its own colours and spent enormous amounts of money in anticipation of basking in competition honours. Successful riders were as popular and as well known as are modern-day baseball players. In England, where modern horse racing originated, the earliest races on record were those that took place in the weekly Smithfield horse fairs during the reign of Henry II, about 1174 A.D. King John (1199–1216) is known to have had race horses in his royal stables. Later, James I helped establish racing at Epsom and Newmarket early in the seventeenth century, and his grandson Charles II (1660–85) was such a great patron of the sport that he came to be known as the 'father of the British turf'. Although British royalty has, in the subsequent centuries, maintained an active interest in racing, none has shaped its development to such a degree as did Charles II.

Horse racing in America was introduced by the colonists who settled New England in the 1600s. In 1654, when the English invaded New Amsterdam and renamed it New York, the commander of their forces was Colonel Richard Nicolls, who lost no time in setting up the first organized racing on the continent. In about 1665 he laid out a two-mile course on Long Island that he called Newmarket. Despite setbacks caused by the Civil War, racing continued to expand until the latter part of the nineteenth century, when those ubiquitous creatures, 'guardians of public morality', managed to constrain legal gambling to the states of Maryland and Kentucky. The twentieth century has, however, seen horse racing become legal once again throughout the United States.

The 1995 Epsom Derby was an event to remember. Not only was it won in record time (by Lamtarra, a son of the great Nijinsky), but it took place on my birthday, and I managed to back the winner at odds of 14 to 1. Lamtarra was having his first outing of the season and only the second start of his career. Coming from a rails position at Tottenham Corner, Lamtarra angled his way towards the outside in the home straight and made up at least six lengths in the final furlong and a half to overtake Tamure close to home and sweep to victory by one length. He gave jockey Walter Swinburn his third win in the world's most famous flat race, following his winning rides on Shergar (1981) and Shahrastani (1986).

Lamtarra, ridden by Walter Swinburne winning the 1995 Epsom Derby

The full results of the race and the starting odds of each horse are shown in the following table.

Position in race	Name of horse	Odds	Subjective Probability of Winning
1st	Lamtarra	14 to 1	0.067
2nd	Tamure	9 to 1	0.100
3rd	Presenting	12 to 1	0.077
4th	Fahal	50 to 1	0.020
5th	Court of Honour	66 to 1	0.015
6th	Vettori	20 to 1	0.048
7th	Riyadian	16 to 1	0.059
8th	Humbel	25 to 1	0.038
9th	Munwar	8 to 1	0.111
10th	Salmon Ladder	50 to 1	0.020
11th	Pennekamp	11 to 8 (favourite)	0.421
12th	Korambi	150 to 1	0.007
13th	Spectrum	5 to 1	0.167
14th	Daffaq	500 to 1	0.002
15th	Maralinga	200 to 1	0.005

The odds reflect the amount of money bet on a horse and are directly proportional to the confidence the betting population has in that horse. Although this is not necessarily synonymous with the actual winning chance of the horse concerned, the odds can be converted into a subjective probability that the horse will win the race by use of the formulae given at the end of Chapter 2. The results of this calculation for the 1995 Epsom Derby are given in the last column of the table.

Now, assuming that all 15 horses don't fall over, one of them must win. So we might expect that the calculated probabilities of winning would sum to 1. But here, summing the 15 probabilities gives the value 1.155. What has gone wrong? Nothing, except that the bookmakers are ensuring that whatever the odds are on the horses in the race, and whatever the result, they remain odds-on to make a profit. (Bookmakers call this making a book which is 'overround', which means profitable, rather than 'underround' which is a losing proposition. Underround is not a word to use within spitting distance of your normally amiable local bookmaker.)

Basically, the money to be returned to punters (bettors) is discounted and a percentage of the gross bet retained by the bookies. The amount of the discount and the amount retained can both be determined from the sum of the subjective probabilities. The discount is simply 1 over this sum, and the amount retained is one minus the result. So in the case of the 1995 Derby, the discount is $\frac{1}{1.155}$ or 0.86, or 86%—so 14% of the gross bet was retained. To make a profit, a discerning bettor must be at least 15.5% more accurate than the collective wisdom reflected in the betting volume. The bookie's advantage over the punter is beautifully summarized by the Oxford mathematician, J.M Hammersley, in an article in a small book entitled *Chance in Nature*.

> A punter puts his shirt on a horse in a mixture of hope and ignorance, not only the requisite ignorance of whether his fancy will win or loose, but often enough ignorance of the gap between the bookie's odds and the nag's chances. The gap is the bookie's livelihood. Individual punters may win or loose individual bets as the wheel of fortune spins, but in the long run the bookie makes a steady profit from the law of averages, and the longer the run or the surer his book, the greater his profit.

In the 1995 Derby, you would have needed a pair of powerful binoculars to spot the 11 to 8 favourite Pennekamp, as Lamtarra passed the finishing post. It is known, though, that favourites win about 30% of horse races. Just how are the odds in horse races generally related to the horses' finishing position? In an ideal world (ideal for the punter rather than the bookie), horses would finish in the order corresponding to their odds: the favourite first, second favourite second and so on. We all know that this rarely happens exactly but is there any evidence of say a *tendency* for starting

"I can't understand it - I'm having
terrible luck these days!"

odds to be related to finishing position? Remarkably, there is. Two statisticians, Dr. Arthur Hoerl of the University of Delaware and Dr. Herbert Fallin, of the US Army Material Systems Analysis Agency, looked at the results of all the thoroughbred horse races run at Aqueduct and Belmont Park in 1970, a total of 1825 races. (Readers, like the author, may well be wondering why such a project would interest the US Army Material Systems Analysis Agency, whatever that is.)

Drs. Hoerl and Fallin examined the results and starting odds of each race and then calculated the average finishing position of the favourite, the second favourite and so on. A summary of their results for races with between five and twelve entries is given here.

		Order by Starting Odds											
N_1	N_2	1	2	3	4	5	6	7	8	9	10	11	12
5	69	2.1	2.4	2.9	3.4	4.1							
6	181	2.2	2.9	3.2	3.6	4.2	4.9						
7	312	2.8	3.2	3.7	4.0	4.3	4.6	5.4					
8	352	2.8	3.2	3.9	4.2	4.7	5.1	5.7	6.4				
9	283	3.1	3.6	4.1	4.6	5.1	5.3	6.0	6.4	7.1			
10	241	3.1	4.0	4.3	5.1	5.3	5.6	6.2	6.5	7.0	7.9		
11	154	3.8	4.0	4.7	5.2	5.7	5.8	6.3	6.9	7.2	7.8	8.5	
12	233	3.9	4.6	5.1	5.4	6.0	6.2	6.7	7.2	7.6	7.7	8.7	9.1

N_1 = number of entries,
N_2 = number of races.

These figures are amazingly consistent in that for each level of number of entries, the order of the average finishing positions matches exactly the order determined by the odds. But before any readers jump to the conclusion that here lies the route to easy money, a pink Cadillac, a villa in Torremolinos and a season ticket for life to visit Gracelands (I was born in Essex), I should point out that in, for example, twelve horse races, the favourite finishes on average fourth. The fact that this is a higher average finishing position than all the horses with greater odds would be little compensation if you had always backed the favourite. And the same principle applies no matter how many horses are in the race.

Thus, it does appeat that, averaged over a large number of races, ordering horses by their starting odds will match their finishing positions relatively well. Sadly, this knowledge is of little use to the punter trying to make an honest living at the race track. But help may be at hand in the work of two more brave statisticians boldly going where most would fear to tread. Drs. Hutson and Haskell of the Department of Agriculture and Resource Management at the University of Melbourne have investigated the pre-race behaviour of horses as a predictor of race finishing order. Over a 20-month period, they assessed the behaviour and appearance of 867 horses entered in 67 races at the Melbourne race courses, Flemington and Moonee Valley. The intrepid Hutson and Haskell found that 'tail elevation' and 'neck angle with the jockey mounted', did help to predict finishing order. But again before readers are tempted to rush out to their nearest race course with protractor and plumb line, I should stress that both tail elevation and neck angle (with the jockey up!) were less accurate predictors than starting odds and weight carried. And no punter has ever got rich relying on either of these.

Perhaps the best way to end this chapter on 'serious gambling' is with the following appropriately highly moral rant against gambling taken from H.G. Wells' book, *The Work, Wealth and Happiness of Mankind*

> From the prince's baccarat and Monte Carlo's roulette and trente-et-quarante, to the soldier's crown and anchor and the errand boy's pitch and toss, it is a history of stakes lost, relieved by incidents of irrational acquisition. It is a history of landslides in an account book. It is a pattern of slithering cards, dancing dice, spinning roulette wheels, coloured counters and scribbled computations on a background of green baize. It is a world parasitic on the general economic organization-fungoid and aimless, rather than cancerous and destructive, in its character. A stronger, happier organization would reabsorb it or slough it off altogether.

Balls, Birthdays and Coincidences

Mysteries are not necessarily miracles.
Goethe

A fool must now and then be right by chance.
William Cowper

There are two major products that come out of Berkeley: LSD and UNIX. We don't believe this to be a coincidence.
Jeremy S. Anderson

Introduction

Mathematicians interested in probability often appear to others to be rather odd individuals. They are, for example, frequently caught pondering the different number of ways in which objects (usually balls) can be placed in containers (usually urns). One of the more fascinating results of such esoteric contemplation is a formula that gives the probability that each urn contains a single ball when a number of balls (say n) are randomly dropped into the same number of urns.

(1) When n balls are randomly dropped into n urns the probability that each urn contains a single ball is given by

$$\frac{n!}{n^n}$$

(2) So, for example, if n is four, the probability is

$$\frac{4!}{4^4} = \frac{4 \times 3 \times 2 \times 1}{4 \times 4 \times 4 \times 4} = \frac{6}{64} = 0.09$$

Here there is about a 1 in 10 chance that each of the four containers will contain a single ball.

B. Everitt, *Chance Rules*, DOI: 10.1007/978-0-387-77415-2_8,
© Springer Science+Business Media, LLC 2008

Deriving such a formula may create much excitement for mathematicians considering their balls and urns, but what use is it for anybody else? Fortunately the result can be applied to events of more general interest than dropping balls into containers. Consider, for example, tossing a die six times. What is the probability that all six faces turn up in six tosses? The answer can be obtained from the foregoing formula, because getting all six faces of a die in six rolls is equivalent to dropping six balls at random into six urns and getting a single ball in each. The required probability is therefore

$$\frac{6!}{6^6} = 0.015$$

So there is less than a 2% chance of all six faces appearing when a die is thrown six times. (Keen dice rollers might like to investigate this empirically.)

As a further example, consider a city in which seven accidents occur each week. The probability of one accident on each day is

$$\frac{7!}{7^7} = 0.006$$

Because this probability is so small, practically all weeks will contain days with two or more accidents. On average only 1 week every 3 years will have one accident per day.

Not content with limiting their interest to situations involving the same number of balls as urns, mathematicians have also considered what happens when the number of containers is greater than the number of balls.

(1) If n is the number of urns and r the number of balls (with n larger than r), the probability that no urn contains more than a single ball is

$$\frac{n \times (n - 1) \times (n - 2) \times \cdots \times (n - r + 1)}{n^r}$$

(2) So for $n = 10$ urns and $r = 7$ balls the probability is

$$\frac{10 \times 9 \times 8 \times 7 \times 6 \times 5 \times 4}{10^7} = 0.06$$

Again this formula becomes more interesting for non-mathematicians if balls and urns are translated into something a little more appealing. Say, for example, that instead of balls in an urn, we consider seven passengers in a lift that stops at each of ten floors. The probability is 0.06 that no two passengers leave at the same floor.

Birthday Surprises

We can also use the formula above to find some perhaps surprising results for birthday matches in a group of people. So let's consider birthdays (not the year) of r people in a year consisting of 365 days. The formula given for the probability that no urn contains more than a single ball can now be applied to find the probability that all r birthdays are different.

The probability that the birthdays of r people are all different is given by

$$\frac{365 \times 364 \times 363 \times \cdots \times (365 - r + 1)}{365^r}$$

This formula looks forbidding, but is easy to evaluate on a computer for various values of r. Here are some examples:

r	Probability that all r birthdays are different
2	0.997
5	0.973
10	0.883
20	0.589
23	0.493
30	0.294
50	0.030
100	0.00000031

You might be asking yourself why the value twenty-three is included in this table? The reason is that it corresponds to a probability of just under a half (0.493) that none of the people have a birthday in common. Hence the probability that *at least two* of the twenty-three people *share* a birthday is a little more than a half (0.507). Because most school classes in the United Kingdom have well over twenty-three pupils, the result implies that in over half of all U.K. classrooms there will be at least two children with the same birthday.

Most people when asked to make a guess how many people are needed in a room to achieve greater than a 50% chance of at least two of them sharing a birthday, put the figure much higher than twenty-three.

Occasionally, misunderstanding of the 'birthday problem' leads to argument and confusion. For instance, the problem was once introduced by a guest on the 'Tonight Show'. The host, Johnny Carson, could not believe the answer of twenty-three and so polled his studio audience of about two hundred people, looking for a match for his *own* birthday. He found none. The guest was unable to explain to Mr. Carson what had gone wrong and how he had, in fact, misunderstood the problem posed and solicited answers to the wrong question. In fact, to have a better than even chance of matching his own birthday, Johhny would have needed 253 people. For those readers who like a little recreational mathematics, this figure is arrived at as follows:

(1) First the chance of any particular person having a birthday *different* from your own is

$$\text{Pr(person having birthday different from your's)} = \frac{364}{365}$$

(2) Thus, the probability that when, say, n people are asked they *all* have a different birthday from yours is

$$\left(\frac{364}{365}\right)^n$$

(3) Finally, the probability that at least one of the n people have the same birthday as you is

$$1 - \left(\frac{364}{365}\right)^n$$

The formula is easily evaluated on a computer for various values of n (see the accompanying table).

n	Probability of at least one birthday the same as yours
1	0.003
5	0.014
10	0.027
50	0.128
100	0.240
200	0.422
253	0.500

The finding that in a group of only 23 people there is greater than a 50% chance of two of them sharing a birthday, comes as a surprise to most people. People do not expect that a matching pair would be so likely in such a small group. But, if a matching pair of birthdays is relatively easy to find, matching triplets are far harder to observe. Without going into the details of the mathematics, which are tricky, it turns out that in a group of 23 people the probability of at least one triplet of matching birthdays is only 0.013, and we have to have a group of 88 people before the probability becomes larger than 0.5.

And what about pairs with nearly the same birthday? Taking 'nearly the same' to mean within one calendar day of each other, then it can be shown that with five people in a group the probability is 0.08, with ten people it is 0.315 and for 14 people it is 0.537.

An interesting variation of the birthday problem involves married couples attending a party. How many couples do we need to have above a 50% chance that there are at least two husband-wife pairs such that the two husbands have the same birthday and so do their wives? Here a rather large gathering with 431 couples is needed.

Finally, we might be interested in determining how many people are needed in a group so that the probability that everyone shares his or her birthday with someone else in the group is greater than 0.5? The answer here is 3064.

(For those readers who would like to know more about the mathematics behind some of these birthday surprises I would again recommend the article mentioned in Chapter 4, *Coincidences and Patterns in Sequences* by Anirban Dasgupta in the *Encyclopedia of Statistical Sciences* published by Wiley.)

Coincidences

Birthday matches in a group of people is a simple example of a coincidence, a surprising concurrence of events, perceived by some as meaningfully related, with no causal connection. Google 'coincidences' and you will be able to spend several happy hours being amazed by 'stories that defy the laws of probability'. Let's begin by looking at some examples.

Golf is one of those games, like cricket, that seems to spawn thousands of fond anecdotes about players and events. Many such tales concern that 'Holy Grail' of golfers, the hole-in-one. Any player who performs this feat is required to buy a large round of drinks upon returning to the club house (in Japan the celebrations are usually far more lavish, and some optimistic players even take out insurance policies to cover their expenses should the miraculous event occur). Such responses to a hole-in-one indicate how unexpected it is. But, in 1989, during the U.S. Open, four golfers, all using seven irons, scored holes-in-one on the sixth hole of the course, in the space of less than two hours. According to the *Boston Globe*;

The odds? A dissertation may be needed in the mathematical journals, but it seems the odds may be 1 in 10 million (from a University of Rochester mathematics professor), to 1 in 1,890,000,000,000,000 (according to a Harvard mathematics professor), to 1 in 8.7 million (according to the National Hole-in-One Foundation of Dallas), to 1 in 332,000 according to *Golf Digest Calculator*, who added that, statistically, this will not happen again in 190 years.

The United States is also the home of our next example of coincidence. Abraham Lincoln and John Fitzgerald Kennedy were the subjects of two of the most tragic assassinations in American history. They do, however, have far more in common than that:

(1) Abraham Lincoln was elected to congress in 1846. John F. Kennedy was elected to congress in 1946.

(2) Lincoln was elected President in 1860. Kennedy was elected in 1960.

(3) Both were very active in civil rights for Black Americans.

(4) Both were assassinated on a Friday, in the presence of their wives.

(5) Each of their wives had lost a son while living at the White House.

(6) Both presidents were killed by a bullet to the head, shot from behind.

(7) Lincoln was killed in Ford's theatre, and Kennedy was killed in a Lincoln made by the Ford Motor Company.

(8) Both were succeeded by vice-presidents named Johnson, both of whom were Southern Democrats, and former senators.

(9) Andrew Johnson was born in 1808, and Lydon Johnson was born in 1908.

(10) Lincoln's assassin, John Wilkes Booth was born in 1839, and Kennedy's killer, Lee Harvey Oswald was born in 1939.

(12) Both assassins were extremists from the South.

(13) Both assassins were themselves killed before they could go to trial.

(14) Booth shot Lincoln in a theatre and fled to a barn. Oswald shot Kennedy from a warehouse and fled to a theatre.

(15) The names Lincoln and Kennedy each have seven letters.

(16) The names Andrew Johnson and Lyndon Johnson each have 13 letters.

(17) Both the assassins were known by three names. Both names, John Wilkes Booth and Lee Harvey Oswald, have 15 letters.

Amazing!

Now a story from the mathematician, Warren Weaver, an expert on the theory of probability.

My next-door neighbour, Mr. George D. Bryson, was making a business trip some years ago from St. Louis to New York. Since this involved weekend travel and he was in no hurry, since he had never been in Louisville, Kentucky, since he was interested in seeing the town, and since his train went through Louisville, he asked the conductor, after he had boarded the train, whether he might have a stopover at Louisville.

This was possible, and on arrival at Louisville he inquired at the station for the leading hotel. He accordingly went to the Brown Hotel and registered. And then, just for a lark, he stepped up to the mail desk and asked if there was any mail for him.

The girl calmly handed him a letter addressed to 'Mr George D. Bryson, Room 307', that being the number of the room to which he had just been assigned.

It turned out that the preceeding resident of Room 307 was another George D. Bryson, who was associated with an insurance company in Montreal but came originally from North Carolina. The two Mr Brysons eventually met, so each could pinch the other to be sure he was real.

Finally, we must note the seemingly incredible story of Margo Adams who won the New Jersey Lottery twice in four months when the chances of a particular person buying exactly two tickets for the lottery and winning with both are only 1 in 17 trillion.

Coincidences to which a self-styled 'expert' attaches very low probabilities are, along with UFOs, corn circles and weeping statues of the Virgin Mary, old favourites of the tabloid press. But it is not only *National Enquirer* and *Sun* readers who are fascinated by such occurrences. Even the likes of Arthur Koestler and Carl Jung have given coincidences serious consideration. Jung even introduced the term *synchronicity* for what he saw as an *acausal* connecting principle needed to explain coincidences, arguing that such events occur far more frequently than can be explained by chance. And Robert Temple, he of the Sirius Mystery (concerned with claiming that aliens from the star Sirius visited the earth some 5000 years ago), is scathing about anybody who tries to find rational explanations for coincidences.

> In my opinion, a mind is healthy when it can perform symbolic acts within mental frameworks which are not immediately obvious. A mind is diseased when it no longer comprehends this kind of linkage and refuses to acknowledge any basis for such symbolic thinking. The twentieth century specializes in producing diseased minds of the type I refer to—minds which uniquely combine ignorance and arrogance. The twentieth century's hard core hyper rationalist would deride a theory of correspondences (*coincidences*) in daily life and ritual as 'primitive superstition'. However, the rationalists' comment is not one upon symbolic thinking but upon himself acting as a label to define him as one of the walking dead [Italics mine].

But Koestler, Jung and Temple get little support from perhaps the greatest twentieth-century statistician, Ronald Alymer Fisher. Fisher, mathematician, rationalist, and clearly somebody Mr. Temple would regard as one of the walking dead, suggests that 'the one chance in a million' will undoubtedly occur, with no less and no more than its appropriate frequency, however surprised we may be that it should occur to *us*.' (See more about Fisher in Chapter 12.)

Most statisticians would agree with Fisher (which is a wise policy since he was almost never wrong) and put down coincidences to the 'law' of truly large numbers. With a large enough sample, any outrageous thing is likely to happen. If, for example, we use a benchmark of 'one in a million' to define a surprising coincidence event, then in the United States with its population of 250 million, we would expect 250 coincidences a day and close to 100,000 such occurrences a year. Extending this argument from a year to a lifetime and from the population of the United States to that of the world (about 5 billion)means that incredibly remarkable events are almost certain to be observed. If they happen to us or to one of our friends, it will be hard not to see them as mysterious and strange. But the explanation is not synchronicity or something even more exotic; it is just our old friend chance at work again.

Let's take one of the examples given earlier, the '1 in 17 trillion' double lottery winner in New Jersey. The '1 in 17' trillion is actually the correct answer to a not-very-relevant question: What is the chance that if you buy one ticket for exactly two New Jersey state lotteries, both will be winners? But the woman involved, like most participants in the lottery, always purchased multiple tickets and, in fact, greatly increased the number of tickets she bought after her first win. Thus, there is a more relevant question here: 'What is the chance of some person, out of all of the millions and millions of people who buy lottery tickets in the United States, hitting a lottery twice in a lifetime?' Stephen Samuels and George McCabe of the Department of Statistics at Purdue University arrived at the answer to this question. Their calculations led them to declare that the event was practically a sure thing, there being a probability greater than a half of a double winner in 7 years somewhere in the United States. Further calculation by our two intrepid statisticians showed that there is better than a 1 in 30 chance of there being a double winner in a 4-month period—the time between the New Jersey woman's two wins.

And what about the seventeen points linking Lincoln and Kennedy and their assassins? These facts about the four men were probably selected from thousands that were originally considered. Those that did not exhibit some 'interesting' connection were conveniently ignored. Consideration of two other assassinated presidents, William McKinley and James Garfield, shows what the enthusiastic 'coincidence creator' can do:

(1) Both McKinley and Garfield were Republicans.
(2) Both McKinley and Garfield were born and raised in Ohio.
(3) Both McKInley and Garfield have eight letters in their last name.
(4) Like Lincoln and Kennedy, both McKinley and Garfield were elected in years ending with zero (1880 and 1900).
(5) Both McKinley and Garfield had Vice Presidents from New York City.

(6) Both the Vice Presidents wore moustaches.
(7) Both McKinley and Garfield were assassinated during the first September of their respective terms.
(8) The assassins of Mckinley and Garfield, Charles Guiteau and Leon Czolgosz had foreign-sounding names.

Coincidences are not always what they're cracked up to be. Indeed according to the mathematician, John Paulos, 'The most astonishing coincidence imaginable would be the complete absence of all coincidences'.

Between the double-six throw of two dice and the perfect deal at bridge, is a range of more or less improbable events that do sometimes happen—individuals *are* struck by lightning, *do* win a big prize in the football pools, and *do* sometimes shoot a hole-in-one during a round of golf. Somewhere in this range of probabilties are those coincidences that give us an eerie spine-tingling feeling, like dreaming of a particular relative for the first time in years and then waking up to find that this person died in the night. Such coincidences are often regarded as weird when they happen to us or to a close friend. However, these events are not weird but merely rare, and explanations other than the simple operation of the laws of chance are hardly ever needed.

Conditional Probability and the Reverend Thomas Bayes

Introduction

Suppose you have two dice and are idly speculating on the chances of rolling one and getting a six and then rolling the other and finding the result is an even number. Because the result of the second throw in no way depends on the result of the first, the two events are independent. Thus, as we have seen in Chapter 2, the probability of both happening is simply the product of the probabilities of each separate event—that is,

$$\Pr(\text{six with first die and even number with second}) = \frac{1}{6} \times \frac{3}{6} = \frac{3}{36} = \frac{1}{12}$$

Now let's suppose you turn your attention to a further question: 'What is the probability, when throwing the two dice, that the number obtained on the die thrown second is larger than the number obtained on the die rolled first?' Here the situation is more complicated, because whether or not the die thrown second shows a larger value than the one rolled first depends on the result obtained in the first roll. For example, if the first die gives a six, the probability of getting a larger value with the second die is zero. However, if the first die shows a five, the chance of the second die having a higher value is the probability that a six is thrown—that is, $\frac{1}{6}$. In this case we have to talk about the probability of the second die showing a higher value than the first, *conditional* on what happened on the first roll.

B. Everitt, *Chance Rules*, DOI: 10.1007/978-0-387-77415-2_9,
© Springer Science+Business Media, LLC 2008

Conditional Probability

Familiarity with conditional probability is of considerable value in understanding topics such as screening for disease and DNA profiling, but it is not the most straightforward of concepts and can sometimes lead to conclusions that seem counterintuitive. Conditional probabilities involve *dependent* rather than independent events and the multiplication rule met in the second chapter needs to be amended a trifle.

(1) The new rule for the probability of the joint occurrence of two dependent events A and B is

$$\Pr(A \text{ and } B) = \Pr(A) \times \Pr(B \text{ given } A).$$

(2) $\Pr(B \text{ given } A)$ is usually written as $\Pr(B|A)$.

As a way to begin exploring the idea of conditional probability, we can return to the dice-throwing example of the introduction. We want to evaluate the probability that when two dice are rolled, the score on the second is larger than the score on the first; the problem is that these are *not* independent events. Let's consider how the event can happen by tabulating all possibilities.

First die score	Second die has higher score if result is	Probability that second die has larger score conditional on the first die's score
1	2 or 3 or 4 or 5 or 6	$\frac{1}{6}+\frac{1}{6}+\frac{1}{6}+\frac{1}{6}+\frac{1}{6}=\frac{5}{6}$
2	3 or 4 or 5 or 6	$\frac{1}{6}+\frac{1}{6}+\frac{1}{6}+\frac{1}{6}=\frac{4}{6}$
3	4 or 5 or 6	$\frac{1}{6}+\frac{1}{6}+\frac{1}{6}=\frac{3}{6}$
4	5 or 6	$\frac{1}{6}+\frac{1}{6}=\frac{2}{6}$
5	6	$\frac{1}{6}$
6	Cannot have higher score	0

Applying our new version of the multiplication rule now leads to the following calculations:

(1) Probability that the first die shows a one *and* the second die a larger score is

$$\frac{1}{6} \times \frac{5}{6} = \frac{5}{36}$$

(*Continued*)

(*Continued*)

(2) Probability that the first die shows a two *and* the second die a larger score is

$$\frac{1}{6} \times \frac{4}{6} = \frac{4}{36}$$

(3) Probability that the first die shows a three *and* the second die a larger score is

$$\frac{1}{6} \times \frac{3}{6} = \frac{3}{36}$$

(4) Probability that the first die shows a four *and* the second a larger score is

$$\frac{1}{6} \times \frac{2}{6} = \frac{2}{36}$$

(5) Probability that the first die shows a five *and* the second die a larger score is

$$\frac{1}{6} \times \frac{1}{6} = \frac{1}{36}$$

(6) Probability that first die shows a six *and* the second die a larger score is zero.

The required probability is now simply the sum of the six separate probabilities, which yields the result that the probability of the second of two dice rolled having a higher score than the first as $\frac{15}{36}$. (Those readers with the necessary patience might like to verify this empirically by tossing two dice several hundred times.)

This example nicely, if somewhat tediously, illustrates the concept of conditional probability, but to justify our claim that 'familiarity with conditional probability is of considerable value,' we need to move away from the rather limited realm of rolling dice. As a first step, we shall consider the condition somewhat misleadingly known as *colour blindness*—misleading because absolute colour blindness with no perception of colours is almost unknown. In fact, colour blindness is a defect in colour perception that involves relative insensitivity to red, green or blue, or some combinations of these. Those who exhibit this condition have nearly always inherited it. Colour blindness is generally detected with the help of the *Ishihara multi-dot test*, where people with normal colour vision see one sequence of numbers and those with problems in colour perception see another sequence.

The genetics of colour blindness dictate that it is more common in men than in women. It is usually said that 5 men in 100 and 25 women in 10,000 suffer from the condition. Thus, the probability of being colour blind is conditional on whether we are talking about men or women. Using the figures given, then, we have

(1) Conditional probabilities of colour blindness for men and women:

$$\text{Pr(being colour blind conditional on being male)} = \frac{5}{100}$$

$$\text{Pr (being colour blind conditional on being female)} = \frac{25}{10000}$$

(2) If we assume that Pr(being male) = Pr(being female) = $\frac{1}{2}$, we can find the *unconditional* probability of being colour blind as

$$\frac{1}{2} \times \frac{5}{100} + \frac{1}{2} \times \frac{25}{10000} = 0.026$$

(3) We see that this is *not* equal to either of the conditional probabilities involved.

As a further example of conditional probability, consider the chance of having spots among those people who have measles—that is, the conditional probability that a person has spots, *given* that the person has measles. This is usually written as Pr(having spots|measles). Because most people who have measles also have spots, this probability must be very high; it is probably close to one. But what about the probability of having measles, given that a person has spots? Most people who have spots will not have measles, so this conditional probability is likely to be very low; it is probably close to zero. This illustrates an important fact about conditional probabilities: In general, the probability of some event A conditional on some other event B, will *not* be the same as the conditional probability of B given A.

Colour blindness and measles are useful examples for illustrating conditional probabilities but two real world examples from the recent past, one from the USA and one from Great Britain, will show how the misunderstanding of conditional probabilities can have serious and even, in one of the cases, tragic consequences.

The trial of O.J. Simpson for the murder of Nicole Simpson in 1994 was one of the most high profile cases seen in the United States for many years. During its 134 days, the trial was watched by a huge audience on prime-time television in the States and in many other parts of the world. The trial appeared to divide opinion about the guilt or innocence of O.J. Simpson, largely by ethnic groups. But after only 3 hours' deliberation, and with an estimated televison audience in the USA alone of 150 million, the jury in the case found O.J. not guilty and he was acquitted.

A member of the defence team later appeared on the Larry King show to explain:

'The statistics demonstrate that only one-tenth of one percent of men who abuse their wives go on to murder them. And therefore it's very important for that fact to be put into empirical perspective, and for the jury not to be led to believe that a single instance, or two instances of alleged abuse necessarily mean that the person killed.

Perhaps the fact that most wives, even those in an abusive relationship, are not murdered (which is all the statement made implies) was a relevant fact for the jury, although I can't think why. More relevant is the fact that there was a body—Nicole Simpson had been murdered by somebody. What needs to be considered here is, given that a wife with an abusive partner has been murdered, what is the probability that the murderer is her abuser? It is this conditional probability that is of real importance to the jury since it is the only probability that provides the *relevant* evidence for them to consider. Best estimates of this conditional probability range from 0.5 to 0.8.

In 1999 in the UK, Sally Clark, a solicitor, was found guilty of the murder of her two infant children and sentenced to two life sentences. During the trial one of the expert witnesses on the prosecution team gave the figure of one in 73 million for the chance of two cot deaths in the same family. This figure arose from squaring the best estimate for the probability of a single cot death namely one in 8500 live births. The expert witness assumed that the cot deaths were independent and so arrived at a value for the probability of two cot deaths that is about one-fifth the probability of winning the UK lottery (see Chapter 6). The figure of one in 73 million was never challenged by the defence team during the trial. Although it is difficult to know how this extremely low probability affected the jury's final verdict, it seems very likely that it played some not inconsiderable part. Which is a pity, since the value of one in 73 million is wrong. The figure of one in 73 million for the probability of two cot deaths in the same family is arrived at by a false assumption, independence; cot deaths in the same family are *not* independent. Genetic and environmental factors predispose babies in some families to be at a greater risk of cot death than average and several studies have shown evidence of an increased frequency of cot deaths in families where one such death has already occurred. The conditional probability of a second cot death in a family, given that there has already been one death, is likely to be far less than the value of 1 in 8500 for a single cot death. Consequently the probability of two cot deaths in the same family will be far, far lower than the value of one in 73 million stated in court. The probability will still be small but it is not difficult to argue that even a figure of say 1 in 100,000 would have had far less impact on the jury than the figure they were quoted. And, of course, the defence team should have pointed out that even the one in 73 million is not strictly relevant, since it needs to be compared with the probability of a mother murdering her two small children, an event that may be even less likely than two cots deaths in the family.

In 2000 an appeal against her sentence by Sally Clark was turned down with the appeal judges ruling that the point raised on the misleading prosecution statistics was of minimal significance. In 2003 a second appeal was successful and Sally Clark's conviction for the murder of her two children was quashed. From her arrest to her release her ordeal had lasted more than 5 years.

Sally Clark died of natural causes on the 16th March 2007, aged 42.

Conditional Probabilities and Medical Practice

The concept of conditional probability is fundamental to the understanding of a number of routine medical practices, but unfortunately, it is often *misunderstood* by both patient and doctor. One area where such understanding is required is in screening programs for various diseases. Here, people who are considered at risk of a disease but who do not have any symptoms (they are asymptomatic) are subjected to one or more diagnostic tests that may help in the early identification of the disease. The implicit idea behind such a procedure is that early identification of the disease will lead to it's more effective treatment. To make the remaining discussion more concrete, we will concentrate on screening for breast cancer.

Breast cancer is by far the commonest form of cancer in women, and it affects nearly a million women, worldwide, each year. It is estimated that about one woman in twelve in Great Britain, and about one in eight in the United States, will develop breast cancer. In 1992, there were more than 15,000 deaths from breast cancer in Britain and about 45,000 in the United States. Great Britain has the highest mortality rate from the disease in the world.

About 5% of breast cancers are thought to be genetic; well-established risk factors for breast cancer are: having no children, starting menstruation early, taking large doses of oestrogens, and exposure to radiation.

Clinicians are constantly stressing the importance of self-examination, because several studies have demonstrated that tumour size is substantially less at the time of diagnosis in women who practise such examination.

Breast cancer takes a terrible toll, and any measure that can reduce its incidence, even to a minor degree, is of enormous value. Such a measure is often considered to be mammography, which is a form of X-ray examination used as a screening procedure on groups of women. A mammogram will lead to either a 'positive' or 'negative' result; the former suggesting that breast cancer is a distinct possibility, the latter that it is unlikely. A positive result is usually followed by a biopsy, which is a microscopic examination, by a pathologist, of tissue from the lump.

But mammograms and all similar diagnostic tests are not infallible. Many women with a positive mammogram will turn out *not* to have breast cancer, and some women who have a negative mammogram will actually have the disease. The usefulness of a diagnostic test depends largely on the values of what are generally known as the test's sensitivity and specificity, and both of these are conditional probabilities.

Sensitivity reflects how well the mammogram can identify women with breast cancer. It is defined formally as the probability of the mammogram being positive given that the woman being tested actually has breast cancer:

$$\text{Sensitivity} = \Pr(\text{mammogram positive}|\text{breast cancer})$$

Specificity reflects the ability of the mammogram to eliminate women who do not have breast cancer. It is defined formally as the probability that the mammogram is negative given that the woman being tested does not have breast cancer:

$$\text{Specificity} = \Pr(\text{mammogram negative}|\text{no breast cancer})$$

In the best of all possible worlds, both probabilities would be one. In practice, we have to accept that both the sensitivity and the specificity of the mammogram will be less than one. In other words mistakes will happen. A 'false positive' occurs whenever the mammogram gives a positive result on a woman who does not have breast cancer; a 'false negative' occurs when a woman with breast cancer obtains a negative mammogram. To avoid both types of mistakes the sensitivity and specificity of the mammogram in particular, and diagnostic tests in general, both need to be high.

But, whilst sensitivity and specificity are critical in evaluating the usefulness of a diagnostic test, the main concern for a woman who gets a positive mammogram will be her chance of having breast cancer. Understandably, most women who get a positive result on a mammogram quickly conclude that they have breast cancer or at least that there is a very high probability that they have the disease. But, as we know from earlier examples, the probability of a positive mammogram, given that a woman has breast cancer, is *not* the same as the probability that a woman has breast cancer given that she gets a positive mammogram. The two may be quite different and it is the latter which the woman agonizing about breast cancer really needs to know. Thus, in general terms, what we want to know is how to find the conditional probability of an event A (breast cancer), given an event B (positive mammogram), from the conditional probability of B given A. Exploring this involves a short digression to meet the Reverend Thomas Bayes and to introduce his wonderful theorem.

An Unsung Statistician

Thomas Bayes was the son of a nonconformist minister, Joshua Bayes (1671–1746). The date of Bayes' birth is unknown, but he was probably born in 1701; he died on the 17th April 1761 and was buried in Bunhill Fields in the City of London in the same grave as his father. Little is known about his life except that he received an education for the Ministry and was, for most of his life, an ordained nonconformist minister in Tunbridge Wells. He was known as a skilful mathematician, but none of his works on mathematics was published during his lifetime. Despite not having published any scientific work, Bayes was elected a Fellow of the Royal Society in 1742. The certificate proposing him for election, dated 8th April 1742 reads

> The Revd Thomas Bayes of Tunbridge Wells, Desiring the honour of being selected into this Society, we propose and recommend him as a Gentleman of known merit, well skilled in Geometry and all parts of Mathematical and Philosophical learning, and every way qualified to be a valuable member of the same.

The certificate is signed:

Stanhope	James Burrow
Martin Folkes	Cromwell Mortimer
JohnEames	

Why was Bayes elected since he seems to have no published work that was relevant? And how did he come to know such men as Philip Stanhope, the famous Earl of Chesterfield and Sir James Burrow, Master of the Crown Office and a lawyer of distinction? Perhaps they drank the waters at Tunbridge Wells or listened to Bayes's sermons there.

Today Bayes is remembered for a paper that his friend Richard Price claimed to have found among his possessions after his death. It appeared in the Proceedings of the Royal Society in 1763 and has often been reprinted. It is ironic that this work that assured his fame (at least among statisticians), the posthumously published *An Essay toward solving a Problem in the Doctrine of Chance*, was ignored by his contemporaries and seems to have had little or no impact on the early development of statistics. The work contained the quintessential features of what is now known as Bayes's Theorem, a procedure for the appropriate way to combine evidence, which has had and continues to have a major influence on the development of modern statistics. For the moment, however we shall introduce Bayes's Theorem in the narrower context of making possible the inversion of conditional probabilities. It provides the means of writing the conditional probability of event A given that event B has occurred, in terms of the conditional probability of event B given that A has occurred and the unconditional probabilities of events A and B. Here it becomes necessary to grasp the nettle of introducing a rather forbidding-looking formula.

Bayes's Theorem is;

$$Pr(A|B) = \frac{Pr(B|A) \times Pr(A)}{Pr(B)}$$

Although this may look scary to readers not familiar with mathematical terminology, it is relatively simple to apply in practice. As our first illustration of the application of Bayes's Theorem we shall consider the following problem, which psychologists Daniel Kahneman and Amos Tversky have used in their research.

A cab was involved in a hit and run accident at night. Two cab companies, the Green and the Blue, operate in the city. The following facts are known:

- 85% of the cabs in the city are Green and 15% are Blue.
- A witness identified the cab as Blue. The court tested the reliability of the witness under the same circumstances that existed on the night of the accident and concluded that the witness correctly identified each one of the two colours 80% of the time and failed 20% of the time.

What is the probability that the cab involved in the accident was actually Blue?

Kahneman and Tversky have found that the typical answer is around 80%; we can use Bayes's Theorem to find the correct answer. Equating the given percentages with probabilities and letting A be the event that a cab is Blue and B the event that the witness says he sees a Blue cab, then what we know is

$$\Pr(A) = 0.15, \ \Pr(B|A) = 0.80$$

To apply Bayes's Theorem to get $\Pr(A|B)$—that is, the probability that the cab is Blue, given that the witness says it is blue—we need to calculate the unconditional probability that the witness says he sees a Blue cab—-that is, $\Pr(B)$. Because the witness is not infallible, he will on occasions correctly identify a Blue cab and on others incorrectly identify a Green cab as Blue. Consequently, the required unconditional probability is given by the sum of the probabilities of these two events:

- Cab is Blue and witness correctly identifies the cab as Blue—probability is 0.15×0.80.
- Cab is Green and witness incorrectly identifies the cab as Blue—probability is 0.85×0.20.

Therefore the required unconditional probability of witness saying that he saw a Blue cab is

$$\Pr(B) = 0.15 \times 0.80 + 0.85 \times 0.20 = 0.29$$

We now have all the terms necessary to apply Bayes's Theorem to give,

$$\Pr(B|A) = \frac{0.80 \times 0.15}{0.29} = 0.41$$

Thus, the probability that the cab actually is Blue, given that the witness identifies the cab as Blue, is less that a half. Despite the evidence the eyewitness offers, the hit-and-run cab is more likely to be Green than Blue! The report of the eyewitness has, however, increased the probability that the offending cab is Blue from its value of 0.15 in the absence of any evidence (known as the *prior probability*) to 0.41 (known as the *posterior probability*).

If you have found this account of applying Bayes's theorem heavy going, you might prefer to consider an alternative approach involving the following diagram. The diagram shows what happens to an imagined population of 100 cabs when the probabilities of correct and incorrect identification are applied to the 15 Blue cabs and 85 Green cabs in this population. The diagram shows clearly that 29 Blue cabs are said to have been seen of which

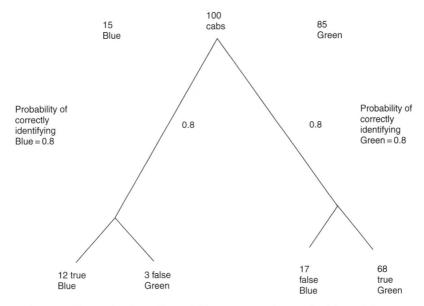

Diagram illustrating how Bayes's Theorem works on the blue and green cab example

12 are actually Blue, giving the probability of a correctly identified Blue cab as 12/29=0.41.

Applying Bayes's Theorem

The same set of steps can be applied to the breast cancer example. Here A will be equated with the event that a woman has breast cancer and B with the event that her mammograph test is positive. The sensitivity of a mammogram, $\Pr(B|A)$, is usually about 0.9 as is its specificity, the probability that the mamograph test is negative given that the woman does not have breast cancer. To use Bayes's Theorem to find the 'reverse' conditional probability that tells us the chance that a woman has breast cancer given that she has a positive mammogram, we need to know the unconditional probabilities of breast cancer and of a positive test from a mammogram. Known figures on breast cancer amongst women who are asymptomatic suggest that the former is about 0.01. What about the probability of a positive test? Such a test can arise from a positive test on a woman with breast cancer or a positive test on a woman *without* breast cancer. The required unconditional positive test probability is obtained from combining the probabilities of each of these—details follow:

(1) The probability of a positive test on a woman with breast cancer is given by

Pr (of a positive test given a woman has breast cancer)

$\times$ Pr (of breast cancer)

(2) This can be written in terms of sensitivity and specificity as

Sensitivity $\times$ Pr(of having breast cancer)

(3) Inserting numerical values gives

Pr (positive test on woman with breast cancer) $= 0.9 \times 0.01$

$= 0.009$

(1) The probability of a positive test on a woman without breast cancer is given by

Pr (of a positive test given woman does not have breast cancer)

$\times$ Pr (of not having breast cancer)

(2) This can be written in terms of sensitivity and specificity as

(1-specificity) $\times$ Pr (woman not having breast cancer)

(3) Inserting numerical values gives

Pr (positive test on woman without breast cancer) $= 0.1 \times 0.99$

$= 0.099$

(1) The *unconditional* probability of a positive test is simply the sum of the probability of a positive test on a woman with breast cancer and the probability of a positive test on a woman without breast cancer.

(2) So from the results above

Pr (of a positive test) $= 0.099 + 0.099 = 0.108$

(3) We can now apply Bayes's Theorem to get the probability that a woman has breast cancer given that she has a positive mammogram:

Pr (woman has breast cancer|positive mammogram)

$$= \frac{0.01 \times 0.9}{0.108} = 0.08$$

A woman who gets a positive mammogram has about a 1 in 12 chance of actually having breast cancer. To put it another way, only about 8 'positive' mammograms out of a hundred are truly positive. Such a low figure clearly has important implications for breast cancer screening and may even suggest that its almost universal acceptance needs to be questioned. The large number of false positives that result from the screening program lead to many unnecessary biopsies and to an unknown number of unnecessary mastectomies. In addition to physical trauma, screening of this kind does psychological harm. For those who have cancer, but who derive no benefit from early diagnosis, screening adds a burden of 'cancer years'—that is, years in which they know that they have the disease and suffer unnecessary anxiety. Undiagnosed, they would have remained symptomless. More serious, because it affects so many more people, is the burden of unnecessary anxiety inflicted by false positive diagnoses. This may well leave wounds to the psyche that heal much more slowly than the wounds of the biopsy.

The acid test of screening programs for breast cancer is, of course, whether or not their introduction reduces breast cancer mortality. At this point, the evidence from clinical trials (see Chapter 12) of screening in the UK is in favour of the procedure, although the decrease in mortality achieved is not huge. Certainly, the apparently blind acceptance by most politicians and many health workers, that screening is ipso facto 'good' for breast cancer and a range of other conditions, is not altogether merited by the scientific studies of screening than have been carried out.

At the risk of becoming tedious, it is worth reiterating the difference between the two conditional probabilities involved in this example:

- *The probability of a positive test given that a woman has breast cancer*, expresses the doubt about whether a woman with breast cancer will test positive. It reflects a characteristic of the mammogram—namely, that it is not infallible.
- *The probability of having cancer given a positive test*, expresses the uncertainty about the cancer when a positive test results.

The difference between the two probabilities is clearly fundamental and Bayes's Theorem gives a way of finding one given the other. The two conditional probabilities are often confused—a confusion often termed *the Prosecutor's Fallacy* because of its frequent appearance in legal cases. Indeed, forensic inquiry is an area where conditional probability and Bayes's Theorem are of great importance in evaluating evidence appropriately (as mentioned earlier in the case of the O.J. Simpson trial). And, as by many clinicians in medicine, both concepts are widely misunderstood by the police and even by those most distinguished members of the legal profession, judges.

"SURE, GENETIC FINGERPRINTING CAN PROVE
SOMEONE'S INNOCENCE — BUT NOT IN A
CASE OF STOCK FRAUD."

Back to the Courtroom

Genetic evidence has been used for over 50 years to draw legal inferences about disputed paternity and criminal matters. More recently DNA 'fingerprinting' has been widely hailed as the greatest breakthrough in forensic science of this century. Its seeming ability to establish guilt or innocence has captured the attention and enthusiasm of police departments, politicians, the legal establishment and the general public. But DNA profiling is not yet perfect, and such DNA evidence needs to be considered in terms of probabilities— and the correct probabilities at that. Suppose, for example, that a crime has occurred and biological material (blood, hair, etc.) that is not from the victim is found at the crime scene. The evidentiary sample is sent for DNA profiling. Now suppose that the police identify a suspect on the basis of other evidence. Analysis reveals that the suspect's profile is identical to that from the sample found at the scene of the crime. Let us further suppose that only one in a million people have this particular DNA profile. Contemplation of such a low figure usually leads the police, tabloid journalists, and (most significantly) jurors and judges to the conclusion that the suspect is clearly guilty. But such a conclusion may often *not* be merited; the explanation involves conditional probability and Bayes's Theorem. First the 'one in a million' figure gives a conditional probability—the probability of obtaining the observed DNA match given that the suspect is innocent. But of far more relevance (particularly to the suspect) is the 'reverse' conditional probability—the probability that the suspect is innocent, given the observed DNA match. The two probabilities may be quite different. Using Bayes's Theorem we know that

$$\text{Pr(innocent|DNA match)} = \text{Pr(DNA match|innocent)}$$
$$\times \frac{\text{Pr(innocent)}}{\text{Pr(match)}}$$

How can we put a figure on the unconditional probability of being innocent? There are a number of possibilities. Recent well-publicised cases suggest that the police tend to go for 'the toss of a coin'—that is, Pr(innocent)=Pr(guilty)=0.5. Using this value we can determine the probability of a match as follows:

(1) The probability of a match is given by
$$\text{Pr(match)} = \text{Pr(match|innocent)} \times \text{Pr(innocent)}+$$
$$\text{Pr(match|guilty)} \times \text{Pr(guilty)}$$

(2) Inserting the various numerical values this leads,
$$\text{Pr(match)} = 0.000001 \times 0.5 + 1 \times 0.5$$

since the guilty person will match.

(3) So from application of Bayes's Theorem we have
$$\text{Pr(innocent|match)} = \frac{0.000001 \times 0.5}{0.000001 \times 0.5 + 1 \times 0.5}$$
$$\approx 0.000001$$

Here the two conditional probabilities are almost equal; the chance of the DNA- matched suspect being innocent is very small. Jurors who find him or her guilty and judges who pass sentence will no doubt congratulate themselves for removing another villain from our streets. But is a figure of a half for the unconditional probability of being innocent really justified? Clearly it is not, as long as the oft-repeated mantra of 'innocent until proven guilty' still holds some claim to be the basis of the legal systems of most countries. It could be argued that since there are approximately 4 billion people in the world and presumably only one guilty person for the particular crime in question, that the required unconditional probability of innocence is extremely large (at least in the absence of any other evidence linking the suspect to the crime). It seems unreasonable however, to assign a person who lives in say Beijing, the same probability of innocence as somebody living close to the scene of the crime. For the former a sensible probability value would be one, for the latter somewhat less than one, based perhaps on the number of people living within a 5-mile radius of where the crime took place. Let's try a value of 0.999999 to see what happens.

$$\Pr(\text{innocent}|\text{match}) = \frac{0.000001 \times 0.999999}{0.000001 \times 0.99999 + 1 \times 0.000001} \approx 0.5$$

There is now a 50% chance that the DNA matched suspect is actually innocent, despite the 'one in a million' chance proclaimed by tabloid headline writers.

Bayes's Theorem can be used in a more general fashion to combine evidence from different sources—it is in fact the optimal way to do this. Even so, when the Court of Appeal in the UK in 1997 considered the role of probability and statistics in assessing the weight of evidence in cases involving DNA profiling, it issued the following statement:

> Introducing Bayes Theorem, or any similar method, into a criminal trial plunges the jury into inappropriate and unnecessary realms of complexity, deflecting them from their proper task. Reliance on evidence of this kind is a recipe for confusion, misunderstanding and misjudgement.

Clearly, the distinguished judges of the Court of Appeal considered coming to a correct decision with a high probability not a 'proper task' of the jury'. Or perhaps it would be more charitable to say that the statement the court issued exposes a weakness in their understanding of the implications of *not* using Bayes's Theorem. Perhaps their Lordships should be asked about what other aspects of mathematics and science a court of law can dismiss as a 'recipe for confusion, misunderstanding and misjudgement'. Should the Pythagorean Theorem or Ohm's Law be equally open to rejection? Such ill-judged comments resulted in the following letter being sent from Professor Adrian Smith, then President of The Royal Statistical Society, to *The Times*.

> Is there no limit to the monumental arrogance of the legal profession?
> Empirical studies have repeatedly shown that human beings make funda-mental errors of logic when trying to combine pieces of evidence in order to evaluate uncertainties. So-called common sense can be shown to be woefully inadequate with issues involving conditional probability. People confuse the probability of an effect given a cause with that of a cause given an effect, and

they overlook base-rates when calibrating the effects of new evidence. In even quite simple situations, there is a proven need for adherence to formal statistical reasoning if gross errors are to be avoided.

And yet, under the heading 'Juries do not apply mathematical formulae' you report a judgement of the Court of Appeal, Criminal Division, which concludes that: Evidence of ... statistical method of analysis in a criminal trial plunged the jury into inappropriate and unnecessary realms of theory and complexity, deflecting them from their proper task... their Lordships... had very grave doubts as to whether that evidence was properly admissible because it trespassed on an area peculiarly and exclusively within the jury's province, namely the way in which they evaluated the relationship between one piece of evidence and another.

So there we have it. Disciplined, scientific reasoning is banned in court!

Perhaps presenting the probability estimates made by ordinary people for the simple Blue and Green cab example discussed earlier might convince lawyers and judges that jurors rely too heavily on some evidence (the eyewitness's report) while underrating more general evidence (the base rates of Blue and Green cabs in the city). To make the example more cogent, they might try substituting White for Green, Black for Blue, persons for cab and mugging for hit-and-run accident.

I cannot leave Bayes's theorem without a passing mention of its use in making a case for the existence of God in the book *The Probability of God: A Simple Calculation that Proves the Ultimate Truth*, by Stephen Unwin, a risk management consultant. Unwin's book is a seductively entertaining and witty attempt to argue that, rather than being a theological issue, the question of God's existence is simply a matter of statistics. Unwin begins by giving the existence and non-existence of God each a 50% prior probability. He then introduces his own subjective conditional probabilities for things like a sense of goodness, religious experience etc., conditional on the existence and non-existence of God, applies Bayes's theorem and comes up with a figure for the probability of God of 67%. And then, despite arriving at the 67% figure after the elaborate application of Bayes's theorem, Unwin tells us that his own personal probability that God exists is 95%, which is a little strange. Unwin's book is a good read but is scientifically fatuous. His prior probability for the existence of God, for example, is not the same as mine, and neither are his subjective conditional probabilities. By changing the figures one could no doubt arrive at almost any posterior probability for the existence of God, although the question of which of the many gods supported by different religions the probability applies to, remains to be asked. In the end even Unwin admits there is no real evidence for the existence of God and falls back, without any apology, on that old standby, faith.

Of course, the real people to ask about the probability of the existence of God are the bookies. In an interview in *The Guardian* in 2007, Graham Sharp, media relations director at William Hills, said there were technical problems with giving odds on the existence of God; 'The problem is', said Mr Sharp, 'how do you confirm the existence of God. With the Loch Ness monster we require confirmation from the Natural History Museum to pay out, but who are we going to ask about the existence of God? The church would definitely confirm his existence.'

Mr. Sharp said William Hill does take bets on the second coming, for which the odds currently stand at 1000 to 1 against. For this confirmation is needed from the Archbishop of Canterbury. But Mr. Sharp is clear that from the bookies perspective, the second coming wouldn't necessarily indicate the existence of God. As ever, bookies like to hedge their bets!

Puzzling Probabilities

I can see looming ahead one of those terrible exercises in probability where six men have white hats and six men have black hats and you have to work out by mathematics how likely it is that the hats will get mixed up and in what proportion. It you start thinking about things like that, you would go round the bend. Let me assure you of that!
Agatha Christie, *The Mirror Cracked*

After all, what was a paradox but a statement of the obvious so as to make it sound untrue?.
Ronald Knox

No question is so difficult to answer as that to which the answer is obvious.
George Bernard Shaw

Introduction

The previous chapter on conditional probability dealt with some serious issues. Chapter 11 which deals with risk assessment, will also have a serious side. In this chapter, as a change of pace, we shall explore some more entertaining and often paradoxical probability problems.

A New Neighbour

(1) A new neighbour arrives at your door to borrow a cup of sugar. You ask whether she has any children. She replies that she has two. 'Any boys', you ask? 'Yes', she says, as she takes her sugar and leaves.

(2) A new neighbour arrives at your door to borrow a cup of sugar. You ask whether she has any children. 'Yes', she replies, 'one five years old and

one who is nine'. 'Is the older a boy', you ask? 'Yes', she says, as she takes her sugar and leaves.

(3) A new neighbour arrives at your door to borrow a cup of sugar. You ask whether she has any children. She replies that she has two. 'Any boys', you ask? 'Yes', she says, as she takes her sugar and leaves. Next day you see her with a small boy. 'Is this your son', you ask? She replies that it is.

In each case the question you ask yourself is 'What is the probability that both her children are boys?'

Consider first situation number 1. Because the two sexes are about equally probable, and the sexes of any two children (born separately) are independent, you might argue that knowing that one of the new neighbour's children is a boy does not affect the probability that the other is a boy, which therefore remains at a half. You would be wrong! To get to the correct answer, we can apply Bayes's theorem.

(1) What we need to find is the following *conditional* probability,

Pr(neighbour's children are both boys|at least one is a boy)

(2) Applying Bayes' theorem we can express this as,

$$\frac{\text{Pr(at least one boy|both children are boys) Pr (both are boys)}}{\text{Pr(at least one boy)}}$$

(3) The first term in the numerator is clearly one, with the second term being $\frac{1}{2} \times \frac{1}{2} = \frac{1}{4}$.

(4) The probability in the denominator is simply:

$$\text{Pr(boy − girl)} + \text{Pr(girl − boy)} + \text{Pr(boy − boy)} = \frac{1}{4} + \frac{1}{4} + \frac{1}{4} = \frac{3}{4}$$

(5) Therefore the conditional probability we require is

$$\frac{1 \times \frac{1}{4}}{\frac{3}{4}} = \frac{1}{3}$$

(6) The probability that the neighbour's other child is a boy is *not* $\frac{1}{2}$ but $\frac{1}{3}$.

(Readers who are not handy with Bayes's theorem can get at this answer by considering that the birth of two children can result in four outcomes: girl-girl, girl-boy, boy-girl, boy-boy. Because we know that there is at least one boy, the first situation is impossible, so the boy-boy outcome is one of three equally likely outcomes and consequently has probability 1/3.)

Now consider the second coming, so to speak, of the new neighbour. Here given the independence of the sexes of any two children born separately, the probability that the neighbour's youngest child is a boy is simply 1/2. The event is independent of the sex of the older child. Here, knowing that the oldest child is a boy, eliminates both the girl-girl and boy-girl outcomes, leaving the boy-boy outcome as one of two equally likely outcomes with corresponding probability of 1/2.

Reasonably straightforward so far, I hope. But now consider the third situation, in which, after your initial meeting with the neighbour, you meet her accompanied by her son in the street. You could be forgiven for thinking that the required probability must be the same as in the first situation, because before spotting the neighbour and her son in the street, you already knew that one of her children was a boy, and you learn nothing new about birth order at the second meeting. Once again, though, you would be wrong. The original question now becomes 'What is the probability that the child you don't see is a boy?' This probability is the same as the probability of the birth of a boy, 1/2. Like the author you may need some time to think about this!

The Fabulous Prize

Now consider another problem. You get to participate in a television game show (I hope that most readers would rather stick pins in their eyes than submit to this ultimate humiliation but bear with me and use your imagination). You are given the choice of three doors, behind one of which is a 'fabulous prize'—house, car, washing machine, world cruise—that sort of thing. Behind each of the other two doors is the recently published autobiography of the show's host. You select a door, and then the host's tall, willowy assistant, Wanda, who knows what's behind the other doors, opens one of the two remaining doors revealing her boss's autobiography. You are now asked whether you would like to switch your choice to the other door not opened by Wanda. What do you do? If you want to maximize your chance of avoiding going home with a tedious 500-page account of the host's tedious life, you should switch. Switching *doubles* your chance of winning the world cruise or whatever. Here is the explanation:

There are three possible places the fabulous prize can be. For argument's sake, let's say you always start by picking door number one. Your chance of winning the prize and avoiding the autobiography is one-third. The probability that the prize is behind door two or three is two-thirds. When the prize *is* behind one of these two doors, the information given by Wanda lets you know which one. (We'll assume that Wanda simply selects one of her two doors at random when the prize is behind neither.) Thus you now have a probability of two-thirds of choosing the door with the prize. By switching you will only take home the book if the prize happens to be behind door number one.

For those still puzzled here is a further, more elaborate explanation. There are two possible choices—switch or do not switch. If you choose not to switch, the possible outcomes are;

Prize is behind	You pick	Wanda opens	Result
Door one	Door one	Door two or three	Stay with one, win
Door two	Door one	Door three	Stay with one, lose
Door three	Door one	Door two	Stay with one , lose

Hence you will win the fabulous prize with probability 1/3 by sticking with your original choice. But if you choose to switch, the outcomes become

Prize is behind	You pick	Wanda opens	Result
Door one	Door one	Door two	Switch to door 3, lose
Door two	Door one	Door three	Switch to door two, win
Door three	Door one	Door two	Switch to three, win

Switching increases the probability that you will win the fabulous prize to 2/3.

If you remain unconvinced you are in good company. When the problem was considered in the 'Ask Marilyn' column of *Parade* magazine in 1990 (a column by Marilyn vos Savant, who claims to have the world's highest IQ, 228), and she correctly recommended switching, she received dozens of letters from mathematicians who maintained that the odds were only fifty-fifty, not 2/3, in favour of such a strategy. One epistle from Dr. E. Ray Bobo will suffice to illustrate what Ms. Savant found in her mailbag.

> You are utterly incorrect about the game-show question, and I hope this controversy will call some public attention to the serious national crisis in mathematical education. If you can admit your error, you will have contributed constructively toward the solution to a deplorable situation. How many irate mathematicians are needed to get you to change your mind?

Poor Dr. Bobo is probably still trying to remove the egg from his face!

More suprising than even Dr. Bobo's failure to appreciate the correct solution is that one of the most famous mathematicians of the twentieth century, Paul Erdös, was also convinced that switching made no difference, as the following report of a discussion with a colleague, taken from Paul Hoffman's fascinating biography of Erdös, *The Man Who Loved Only Numbers*, makes clear.

> 'I told Erdös that the answer was to switch', said Vázsonyi, 'and fully expected to move to the next subject. But Erdös, to my surprise, said, 'No, that is impossible. It should make no difference.' At this point I was sorry I brought up the problem, because it was my experience that people get excited and emotional about the answer, and I end up with an unpleasant situation. But there was no way to bow out, so I showed him the decision tree solution I used in my undergraduate Quantitative Techniques of Management course'. Vázonyi wrote out a decision tree, not unlike the table of possible outcomes that vos Savant had written out, but this did not convince him. 'It was hopeless', Vázsonyi said. 'I told this to Erdös and walked away. An hour later he came back to me really irritated. 'You are not telling me *why* to switch', he said. 'What is the matter with you?' I said I was sorry, but that I didn't really know why and that only the decision tree analysis convinced me. He got even more upset.' Vázsonyi had seen this reaction before, in his students, but he hardly expected it from the most prolific mathematician of the twentieth century.

Vázsonyi only eventually managed to persuade Erdös that switching was indeed the correct strategy by simulating the game on his PC.

Theres No Such Thing as a Surprise Fire Drill

Many organizations like to run 'surprise' fire drills to ensure that their employees know exactly how to respond in the event of a real fire. Often, employees are told that there is to be a surprise fire drill 'one working day next week'. Now, the fire drill cannot be held on Friday, for on Thursday evening the employees would known that because it had not happened already, it was going to happen tomorrow and therefore would not be a surprise. Similarly it cannot happen on Thursday, because Friday having been ruled out, it would be known by Wednesday evening that it was going to take place on the morrow. And so on. *Surprise fire drills are impossible!*

The solution to this seeming paradox involves conditional probabilities.

(1) Assuming the previous week contains no information about when the fire alarm practice will occur, the probability of the practice taking place on Monday is 0.2.

(Continued)

(*Continued*)

(2) On Monday evening, the practice not having occurred, we can say that the probability that the practice will take place on Tuesday *given* that it did not occur on Monday is 0.25.

(3) Similarly the probability that the practice takes place on Wednesday given that it has not been held on either Monday or Tuesday is 0.33, and the probability for Thursday given that no practice took place on Monday, Tuesday and Wednesday is 0.5.

(4) On Thursday evening the conditional probability for Friday given no practice earlier in the week reaches one and the element of surprise has now genuinely vanished.

The changing conditional probabilities throughout the week reflect the changing *degree* of surprise.

Will You Get the Correct Seat?

Imagine an orderly queue of 100 airline passengers waiting to board their flight on which there are 100 seats; imagine further that you are passenger number 100. Each person in the queue has a ticket with a specified seat number. Sadly, the man at the head of the queue has only ever flown before on Easy Jet and so when he boards the plane he selects a seat at random. Subsequent passengers choose their allocated seat if possible, but if it is occupied they choose a seat at random (they are all English, of course). The question is, 'what is your probability of getting the correct seat?' There are various ways to go about finding the answer but probably the most straightforward is to simply consider some simpler situations, say, when there are 2, 3 and 4 passengers (and 2, 3 and 4 seats). So we have;

Two passengers, A and B, with allocated seats S1 and S2,

$$
\begin{array}{ccc}
\text{Possibility one}: & \text{S1} & \text{S2} \\
 & \text{A} & \text{B} \\
\text{Possibility two}: & \text{S1} & \text{S2} \\
 & \text{B} & \text{A}
\end{array}
$$

Probability of the second passenger, B, getting the correct seat S2 is ½.

Three passengers, A, B and C with allocated seats S1, S2, and S3,

Possibility 1 :   S1   S2   S3
                  A   B   C

Possibility 2 :   S1   S2   S3
                  B   A   C

Possibility 3 :   S1   S2   S3
                  C   A   B

Possibility 4 :   S1   S2   S3
                  C   B   A

Probability of the third passenger, C, getting the correct seat S3 is again ½.

Four passengers, A, B, C and D with allocated seats S1, S2, S3 and S4,

Possibility 1 :   S1   S2   S3   S4
                  A   B   C   D

Possibility 2 :   S1   S2   S3   S4
                  B   A   C   D

Possibility 3 :   S1   S2   S3   S4
                  C   A   B   D

Possibility 4 :   S1   S2   S3   S4
                  D   A   B   C

Possibility 5 :   S1   S2   S3   S4
                  D   A   C   B

Possibility 6 :   S1   S2   S3   S4
                  C   B   A   D

Possibility 7 :   S1   S2   S3   S4
                  D   B   A   C

Possibility 8 :   S1   S2   S3   S4
                  D   B   C   A

Probability of fourth passenger, D, getting the correct seat, S4, is once again ½.

Looking at these three examples, we see that the last passenger will only ever have a choice of seat S1 or his correct seat and his probability of each is simply a half. And this will be true with any number of passengers where the number of passengers and the number of seats is the same. Surprising!

Finding the Right Hat

Imagine you go for a meal at an exclusive restaurant and check in your hat. At the end of the meal, you and a number of other diners—say, five people in total, go to reclaim your hats. The rather bored hatcheck person hands out the relevant five hats at random. On average how many hats will be returned to their rightful owners, and what would you guess is the probability that at least one person gets the correct hat? And how would the average and your guess change if fifty people were involved or one hundred or ten thousand? In fact, the average remains the same and takes the value 1 however many hats are involved, and the probability is also practically independent of the number of hats being randomly handed out and is roughly $\frac{2}{3}$. Some *exact* values for particular number of matches follow:

Probabilities of *m* hats being given to their rightful owners When *N* hats are handed out at random

m	N = 3	N = 4	N = 5	N = 6	N = 7
0	0.333	0.375	0.367	0.368	0.368
1	0.500	0.333	0.375	0.367	0.368
2	—	0.250	0.167	0.187	0.184
3	0.167	—	0.083	0.056	0.061
4	—	0.042	—	0.021	0.015
5	—	—	0.008	—	0.003

The probability of at least one match is simply one minus the probability of no matches, giving in each case a value of about 2/3. (The gaps reflect the impossibility of having $N - 1$ correct hats without having all N correct.)

This problem can, of course, be formulated in a number of different ways. For example, say two equivalent decks of cards are shuffled and matched against each other. The same card occupying the same place in both decks is equivalent to receiving the correct hat. Or we might consider a rather scatterbrained secretary placing letters at random into previously addressed envelopes. Tables of probabilities such as given for this problem are useful in testing guessing abilities—for example, in wine tasting or psychic experiments. Any actual insight on the part of the subject will appear as a departure from randomness. For example, if you take cards from a well-shuffled deck and ask somebody to name each card just before it is shown, then with random guessing you will see, on average, one card correctly identified; if four or more cards are correctly identified you have either found another Uri Geller, or a genuine psychic.

Records or Why Things Were Better When I Was Young

Do you remember when you were a child? Almost every summer there seemed to be record high temperatures, almost every winter record falls of snow (as noted by Bill Bryson) and nearly every spring less and less rain to put a stop to outside pursuits. Such childhood perceptions are beautifully summarized by the following extract from Bill Bryson's wonderful book, *The Life and Times of the Thunderbolt Kid*;

> Of course, winters in those days as with all winters of childhood, were much longer, much snowier and more frigid than now. We used to get up to eleven feet of snow at a time-we seldom got less, in fact-and weeks of arctic weather so bitter you could pee icicles.

As you get older the weather each year seems much the same as last year and record temperatures etc., seem to happen less and less. Is this simply a psychological phenomenon in which the old feel that everything was better in their youth or is there more to it?

To make the discussion that follows simpler, let's consider the temperature example and stick to say, the average monthly temperature for August for a number of years. We will use the term 'record' for those years which have the largest temperature amongst all temperatures recorded up to the year in question. So, for example, if for 10 years the average August temperatures in degrees Centigrade were

$$23, 20, 22, 25, 24, 21, 28, 27, 26, 30$$

we would say that the values 23, 25, 28 and 30 were records. The first observation is *always* called a record because it is the largest value recorded at that time. (For what follows we have to assume that the values in the series of observations are essentially random so that athletic records, for example, where people systematically try to improve their performance are not included. And we also assume that no two observations in the series are equal.)

If there are n observations in our series (n temperatures, for example) they can be ordered in $n!$ ways, and all orderings have the same probability. The probability that the first observation in the series is a record is clearly one by definition. The probability that the second observation is a record is just $1/2$. The probability that the third observation is a record is $1/3$ and so on until the last (the nth) observation with probability $1/n$. If we have a large number of mutually independent series each of length n and we count the number of records in each of them, then the average of these counts can be shown to approach the value

$$1 + \frac{1}{2} + \frac{1}{3} + \ldots\ldots + \frac{1}{n}$$

We can use this result to find the values of n for which the average number of records in the series is equal to or exceeds particular values. For example, for the average number of records to equal or exceed two we need only four observations in the series. Other values are:

n	4	11	31	83	227	616	1674	4550	12367
N	2	3	4	5	6	7	8	9	10

Where the average number of records in the series is equal to or exceeds N

So returning to the record August temperatures example, we see that a child of age 11 would, on average have seen three record Augusts, a person aged 31 would have seen an average of four and an elderly person of 83 would have seen an average of five. (An intuitive explanation is that as time goes by, and the numbers of records increase, it gets harder and harder to surpass the last record achieved.) Here may be why certain things at least seem to get worse as we get older.

(For a full explanation of records but one that demands some technical expertise I recommend, *Mathematics of Chance* by Jiri Andel, published by Wiley.)

Choosing the Most Bubbly Secretary

Choosing a secretary is at least as important as choosing a spouse. In both cases the wrong choice can lead to disaster and tears for both parties. And it has to be remembered that getting rid of an unsatisfactory spouse may be easier than shedding an unsatisfactory secretary. So how is a manager looking for a new secretary to choose? No doubt the manager will have several criteria in mind that the ideal secretary should satisfy—personality, efficiency, IT skills etc., will all be important. But for simplicity let's suppose that there is a single criterion say, 'bubbleness', for which each candidate can be given a score and high scores represent a more desirable level of bubbleness for the manager. (The bubbleness scale is a visual analogue scale with anchor points at zero for the secretary from hell, to 100, for the secretary who is always happy, supportive, helpful, a shoulder to cry on and who understands that at times your spouse does not understand you—currently the scale is only used covertly.) An advertisement for the job has been answered by n people. Candidates are interviewed one by one, and at the interview the manager rates the candidate's bubbleness looking to give the job to the candidate with the highest score on this scale. The problem is that each applicant wishes to know immediately whether he or she is to be hired, and applicants who are turned down suffer an immediate attack of pique, and leave, never to return. The question is to find a managerial strategy that ensures the highest probability that the most-bubbly applicant is hired. (We shall assume that all candidates have different scores on the chosen scale.)

Let's begin by considering a situation where $n = 3$ and the scores of the three candidates on the bubbleness scale are 10, 20 and 30. There are six possible orders in which the candidates could be seen and we shall assume that each order is equally likely. The scores in each of the six orders are:

$$10, 20, 30$$
$$10, 30, 20$$
$$20, 10, 30$$
$$20, 30, 10$$
$$30, 10, 20$$
$$30, 20, 10$$

If the manager simply chooses the first candidate or the last or any other particular candidate we are led to a probability of choosing the most-bubbly applicant of 1/3. But suppose the manager lets the first candidate leave and chooses the next one with a higher score; this leads to the candidate with the highest score of 30 in the second, third and fourth permutations thus giving a probability of selecting the most desirable candidate of 1/2. Letting two candidates go leads to the correct choice only in the first and third permutations. So, for three candidates the best strategy is to let the first one to be interviewed leave disappointed, and choose the next one with a higher score if this is possible, or otherwise choose the final candidate interviewed.

So it seems that the best strategy is to let a certain number of candidates leave and then take the next one who has a higher score than any of the ones seen thus far. Can we find the probability that the best candidate is selected by such a strategy for different values of n and different values for the number of candidates that are let go (we shall denote this number by m)?

Adopting the strategy outlined above, the manager can fail to choose the bubbliest candidate if (1) the most bubbly candidate is among the first m candidates interviewed or (2) the most bubbly candidate is not among the first m candidates but is preceded in the interviews by a candidate with a higher score than any in the first m. Now consider the probability of the desired candidate being the $(m + 1)$th person interviewed. This is simply the probability that any of the n candidates is the best, namely $1/n$. Again the probability that the $(m + 2)$th candidate interviewed has the highest score is $1/n$ but this candidate will only be selected if the previous highest score amongst the $(m + 1)$ candidates already interviewed is among the first m candidates who are turned away and this has probability $m/(m + 1)$. So the probability of the $(m + 2)$th candidate being the one the manager would like for a secretary is simply the product of these two probabilities to give $m/(n(m + 1))$. Hopefully it is now easy to see the general pattern, so the probability of the $(m + 3)$th candidate being the best, for example, is $m/(n(m + 2))$. Finally the probability that the last candidate to be interviewed is the most bubbly is $m/(n(n - 1))$.

Hence adopting the strategy with n candidates of asking the first m to be interviewed to leave and then choosing the applicant with a score better than any of those seen so far, the probability of choosing the one who is most bubbly is just the sum of all the probabilities from that for candidate $m + 1$ to that for candidate n—that is, the rather daunting expression:

$$\Pr(s, m) = \frac{m}{n} \left(\frac{1}{m} + \frac{1}{m + 1} + \ldots \frac{1}{n - 1} \right)$$

For any given value of n, the number of candidates, we need to find the corresponding value of m that maximizes the expression above. If $n = 100$ the probability is maximal for $m = 37$, so the manager would interview the first 37 candidates and reject them, and then choose the next candidate whose bubbleness score exceeds that of the preceding applicants. The probability that the most bubbly candidate would be chosen by this strategy in this particular case is 0.371. For relatively large values of n it can be shown (the usual expression for 'it requires some mathematics') that the optimal value of m is approximately $(n-1)/e + 1$ where e is the base of natural logarithms and takes the value 2.7183. So for various values of n the optimal values of m are:

n	50	60	70	80	90
m	19	22	26	30	33

The St. Petersburg Paradox

In Chapter 4 we looked at some games involving tossing coins. Let's reintroduce our two players, Peter and Gordon, and consider a further coin tossing game. The game goes like this: A fair coin is tossed—if a head is thrown on the first toss, Peter will give Gordon a dollar; if the first toss is a tail, but a head appears on the second, Peter will give Gordon$2 and so on, with the amount Peter gives to Gordon being doubled each time a tail appears before a head is tossed, and when a head has been tossed the game is over. What should Gordon pay Peter for the privilege of playing the game? To make the game fair Gordon needs to pay an entry fee equal to the average amount he would expect to win in playing the game. What would you think is fair? If you were Gordon would you be happy to pay $25, or $50 or maybe even $100? In fact you should be prepared to mortgage your house, your wife, your children and your dog and offer to pay the largest amount you can raise, since it can be shown mathematically that your average winnings in this game are infinite! But good old common sense (not always to be relied on, of course) suggests it would only be worth investing a modest amount on the game. In fact, way back in the eighteenth century, when Georges Louis Leclerc, Comte de Buffon (1707–1788), made an empirical test of the matter, he found that in 2084 games Peter would have paid Gordon $10,057, giving an average of something less than five dollars; far from paying some huge amount to play, Gordon should offer about this amount.

So what is happening? Well the mathematics are correct and the average amount won is infinite, but the paradox arises because Peter's fortune is necessarily finite and so there is a flaw in the way the game is arranged.

Peter could not pay the unlimited sums that may be required in the case of a long delay in the appearance of a head.

All Probabilities Are Conditional

Suppose the following figures have been collected to throw light on whether there is racial equality in the application of the death penalty in a particular country.

	Death Penalty	Not Death Penalty
White defendant found guilty of murder	190	1410
Black defendant found guilty of murder	170	1490

Using the relative frequency approach to assigning probabilities even though the numbers are relatively small, we find

$$\Pr(\text{white defendant being given death penalty}) = \frac{190}{1600} = 0.12$$

$$\Pr(\text{black defendant being given death penalty}) = \frac{170}{1660} = 0.10$$

The two probabilities are very similar, and we might therefore conclude that there was indeed racial equality in the application of the death penalty. But suppose we now dig a little deeper and introduce the race of the victim. This yields the following revised figures:

White Victim

	Death penalty	Not death penalty
White defendant found guilty of murder	190	1320
Black defendant found guilty of murder	110	520

(Continued)

(*Continued*)

Black Victim

	Death Penalty	Not death penalty
White defendant found guilty of murder	0	90
Black defendant found guilty of murder	60	970

We can now find some relevant *conditional* probabilities:

$$\text{Pr(death penalty}|\text{defendant white and victim white)} = \frac{190}{1510} = 0.13$$

$$\text{Pr(death penalty}|\text{defendant black and victim white)} = \frac{110}{630} = 0.18$$

$$\text{Pr(death penalty}|\text{defendant white and victim black)} = \frac{0}{90} = 0$$

$$\text{Pr(death penalty}|\text{defendant black and victim black)} = \frac{60}{1030} = 0.06$$

Claims of racial equality in the application of the death penalty now look a little more difficult to sustain!

Perhaps the moral behind all these examples (if there is one) is that evaluating probabilities sometimes requires a bit of lateral thinking and sometimes recourse to that theorem of the Reverend Thomas Bayes.

Taking Risks

*You've got to keep fighting—you've got to risk your life every six
months to stay alive.*
Elia Kazan

*One hour of life, crowded to the full with glorious action, and filled
with noble risks is worth whole years of those meanobser vances of
paltry decorum.*
Sir Walter Scott

Introduction

A famous statistician would never travel by airplane, because he had
studied air travel and estimated the probability of there being a bomb on
any given flight was one in a million, and he was not prepared to accept this
level of risk.

One day a colleague met him at a conference far from home and asked,
'How did you get here, by train?'

'No, replied the statistician, 'I flew'.

'What about the possibility of a bomb?' his friend enquired.

'Well I began thinking that if the odds of one bomb are one in a million,
then the chance of there being two bombs is very, very small and is a risk I
am prepared to take. So now I bring along my own bomb!'

Now I know that this is an old and not particularly funny story, but it
does help to introduce the rather curious fashion in which many people
evaluate the risks in their lives. Risk is defined in my dictionary as 'the
possibility of incurring misfortune or loss'; gambling is a voluntary form of
risk-taking. Risk is synonymous with hazard, and almost everything we do
in life carries some element of risk because the world is a very risky place.
When I get out of bed in the morning, for example, turn on the light and
then go downstairs for breakfast, I risk death or serious injury from

B. Everitt, *Chance Rules*, DOI: 10.1007/978-0-387-77415-2_11,
© Springer Science+Business Media, LLC 2008

electrocution and falling. My first cup of tea of the day is loaded with toxins, many of which increase the risk of suffering and perhaps dying from various forms of cancer. The risks increase when I leave the house and travel to work. My car, train or bicycle journey is fraught with potential hazards. While at work, although a non-smoker myself (except in very specific circumstances—see later), I may be bombarded with the tobacco smoke of less thoughtful colleagues. In this way I am exposed to the poison which is the cause of 15% of all deaths in the United Kingdom and the United States. I could go on, but the message is clear: There is no such thing as a no-risk world.

"Typical Jenkins - always afraid to take risks."

Even that simple pleasure of walking to and from the pub in the evening involves a number of risks including:

(1) Being run over
(2) Being mugged
(3) Being involved in a fight
(4) Being arrested by the police for being drunk and disorderly
(5) Being forced to take part in a karaoke competition
(6) Being asked to buy a round of drinks.

The chances of each of these possibilities are considerably different, as are the degrees of misfortune suffered from each (for many of my more sensitive readers, for example, number 5 would probably represent a major trauma). Most people would be willing to ignore all such risks in pursuit of a refreshing pint although making that pint a habit can itself entail some risk). In short, we accept that for the sake of obtaining some pleasures or necessities in life, a certain degree of risk is worth taking.

Quantifying Risk

Quantifying and assessing risk, involves the calculation and comparison of probabilities, although most expressions of risk are compound measures that describe both the probability of harm and its severity. Americans, for example, run a risk of about 1 in 4000 of dying in an automobile accident. The probability is 1 out of 4000 for injuries lethally severe. This projection is derived by simply noting that in the last few years, there have been about 56,000 automobile deaths per year in a population of approximately 224 million. The figure of 1 in 4000 expresses the overall risk to society; the risk to any particular individual clearly depends on their exposure: how much they are on the road, where they drive and in what weather, whether they are psychologically accident-prone, what mechanical condition their vehicle is in, and so on. Because gauging risk is essentially probabilistic, we must remember that a risk estimate can assess the overall chance that an untoward event will occur but is powerless to predict any specific event. This is like knowing that the probability of tossing a head with a fair coin is one-half but being unable to predict which tosses will result in heads and which in tails.

Most people, however are not comfortable with mathematical probability as a guide to living and are not always able to make rational choices about the differing risks faced in undertaking particular activities or indulging in particular lifestyles. Many people, for example, are afraid to fly with commercial airlines, but practically nobody is afraid of travelling by train or driving in cars—a stunning victory of subjectivity over statistics, as the following figures clearly demonstrate:

Cause of death	Odds
Car trip across USA	1 in 14,000
Train trip across USA	1 in 1,000,000
Airline accident	1 in 10,000,000

One reason for this excessive and clearly misplaced fear of flying is perhaps the general view that dying in a plane crash must be a particularly nightmarish way to die.

Research shows that people tend to overestimate the probability of unfamiliar, catastrophic, and well-publicized events and underestimate the probability of unspectacular or familiar events that claim one victim at a time. An illustration of this is provided by the results of polls of college students and members of the League of Women Voters in Oregon. Both groups considered nuclear power their number-one 'present risk of death,' far ahead of motor vehicle accidents which kill 50,000 Americans each year, cigarette smoking which kills 150,000 or handguns which kill 17,000. Average annual fatalities expected from nuclear power, according to most scientific estimates, are fewer than 10. Nuclear power does not appear to merit its number-one risk rating. The two well-educated and influential segments of the American public polled in Oregon seem to have been misinformed. The culprits are not difficult to identify: Journalists report almost every incident involving radiation. A truck carrying radioactive material is involved in an accident, a radioactive source is temporarily lost, a container leaks radioactivity—all receive nationwide coverage, whereas the three hundred Americans killed each day in other types of accidents are largely ignored. Use of language such as *deadly radiation* and *lethal radioactivity* also biases the reader's assessment of the risks involved. I suspect that the correponding phrases, *deadly motor cars* or *lethal water*, would not sell enough newspapers, although thousands of people are killed each year in road accidents and by drowning. The problem is highlighted by a study carried out over a 2-year period into how frequently different modes of death became front-page stories in the *New York Times*. It was found that the range was from 0.02 stories per 1000 annual deaths from cancer to 138 stories per 1000 annual deaths from airplane crashes.

The problem is that the news media help to shape public discussion surrounding risk issues by how risks and uncertainties are portrayed and represented in their stories. Journalists are largely interested in the newsworthiness of a story to the public and the tendency is for them to present the novel or dramatic over the more common, and often more serious, risks. And the problem is compounded because few journalists have the scientific background required to make sense of the wealth of data that is frequently presented to help in the understanding of risks. According to Roy Greenslade, Professor of Journalism at the City University, London, 'most journalists become journalists because they were good at English and bad at maths, so there is an overall bias against mathematical understanding'.

Several scientists have blamed the media for what they see as public overreaction to risk. The following is a quotation from the physicist Bernard Cohen:

> Journalists have grossly misinformed the American public about the dangers of radiation and of nuclear power with their highly unbalanced treatment and their incorrect or misleading interpretations of scientific information. This misinformation is costing our nation thousands of unnecessary deaths and wasting billions of dollars each year.

Professor Cohen might be guilty of a little sermonizing here, but his view appears to be backed up by the several studies concerning media coverage of the accident at Three Mile Island. Repeated stories about Three Mile Island redirected public attention away from competing news stories including that involving the worst airline accident in American history (at that time) which killed 274 people. Media focus was on Three Mile Island mobilizing fears about risks associated with nuclear power and the possibility of similar nuclear accidents occurring more frequently elsewhere in the country. There are, of course, well documented and persuasive arguments against nuclear energy—safety of reactors, disposal of radioactive waste, and the threat of nuclear terrorism to name but three. But the fact still remains that public concern about the overall risks to life and the environment appears to be out of proportion to concern about other risks in our lives and is not, at present at least, supported by compelling empirical evidence. (The occurrence of several other Chernobyl type disasters in the next decade or so might make it necessary to revise this sentence.)

Contamination is another area where the public's perceptions of risk are distorted. Most chemicals in the water, soil and air present trivial hazards, far less worrying than more mundane risks. Pesticide residues discovered on fruit and vegetables often causes an outcry in the media, but exposure to these residues is probably far less risky than smoking one or two cigarettes a day. In a similar vein are recent, disturbing reports of the adverse effects on people's health of living near overhead power lines although the evidence is highly subjective.

If the risk is very small but easy to avoid without a major change in lifestyle, it is perhaps rational to avoid it. But most people's approach to risk-taking is highly irrational. They refuse to engage in activities that have known, but quite negligible, risks, but they fearlessly participate in activities that pose dangers of greater orders of magnitude. In his excellent *A Mathematician Reads The Newspaper*, John Paulos conjures up the image of the family, travelling by car, mother obese and munching on potato chips, the father resting two fingers on the steering wheel and alternatively sipping from a can of beer and puffing on a cigarette, and a small child standing between them playing with the rearview mirror. The parents' conversation involves their anxieties about the poisons and chemicals around them, the dangers to their child of the power lines near to their house, and their fears about the increase in the use of nuclear power.

More telling than this imagined scene in illustrating the general irrationality of risk-taking, is the sad story related by the late Alaistair Cooke in a *Letter from America* during the Gulf War of 1990. An American family of four cancelled their planned holiday to Europe because of the fears of terrorist attacks on their country's airlines. They decided to drive to San Francisco instead. At the last road junction before their itinerary was to have led beyond their small Midwest town, they collided with a large truck with fatal consequences for them all.

Risk Presentation and Risk Perception

In a classic study of risk perception undertaken in the late 1970s, participants were asked to estimate the mortality rates of various risks (in the United States), including flood, tornados and homicide. The results are shown in the following diagram and reveal that people overestimate rare risks, such as death from botulism or smallpox vaccination, and they underestimate frequent risks, such as cancer or strokes. (The straight line on the diagram shows where actual number of deaths and estimated number of deaths are equal. The curved line shows the observed relationship between actual and estimated number of deaths.)

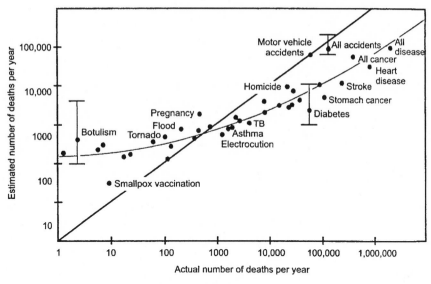

Risk and risk perception

Perception (and, unfortunately, misperception) of risks plays an important role in governmental decision-making. Governments may spend huge amounts of money protecting the public from, say, nuclear radiation, but are unlikely to be so generous in trying to prevent motor vehicle accidents. They react to loudly voiced public concern in the first case and to the lack of it in the second. But if vast sums of money are spent on inconsequential hazards, little will be available to address those that are really significant. The problem is that risks from dramatic or sensational events, such as homicides and natural disasters, tend to be greatly overestimated. Risks from undramatic causes such as asthma and diabetes tend to be underestimated. More rational debate will arise only when there is greater public appreciation of how to evaluate and, in particular, compare risks. Comparisons are generally more meaningful than absolute numbers or probabilities, especially when these absolute values are quite small. And, making such comparisons may help you avoid getting into a panic about the risks of life until you have compared the risks that worry you with those that don't, but perhaps should. Risks are most often presented in the form of probabilities (or their equivalent). The accompanying table, for example, gives the annual fatality rates per 100,000 persons at risk from a wide variety of causes (probabilities can be found by dividing each rate by 100,000).

Risk	Rate
Motorcycling	2000
Smoking	300
Hang-gliding	80
Farming	36
Boating	5
1 Diet drink/day (saccharin)	1
Floods	0.06
Lightning	0.05

Such figures are a start, but they fail to reflect the fact that some hazards (for example, pregnancy and motorcycle accidents) cause death at a much earlier age than others (for example, lung cancer due to smoking). One way to overcome this problem is to calculate the average loss of life expectancy due to exposure to the hazard. The following table gives some examples:

Risk	Days lost
Being unmarried–male	3500
Cigarette smoking–male	2250
Heart disease	2100
Being unmarried–female	1600
Cancer	980
20% Overweight	900
Low socioeconomic status	700
Stroke	520
Motor vehicle accidents	207
Alcohol	130
Suicide	95
Being murdered	90
Drowning	41
Job with radiation exposure	40
Illicit drugs	18
Natural radiation	8
Medical X-rays	7
Coffee	6
Oral contraceptives	5
Diet drinks	2
Reactor accidents	2
Radiation from nuclear industry	0.02

So, for example, the average age at death for unmarried males is 3500 days younger than the corresponding average for men who are married. This implies, in general terms at least, that the institution of marriage is 'good for men'. But be warned, it does not imply a cause (marrying) and effect (living 10 years longer) applicable to every individual. Unfortunately for some men, marrying that childhood sweetheart might be a mistake that leads to an early grave.

Nevertheless, the ordering in this table should largely reflect society's and government's ranking of priorities. Thus, rather than introducing legislation about the nuclear power industry or diet drinks, a rational government should be setting up computer dating services, stressing the advantages of marriage and encouraging people to control their eating habits. It is hard to justify spending any money or effort on reducing radiation hazards or dietary hazards such as saccharin. Pouring resources into publicizing the dangers of illicit drugs, whilst allowing the continuation of the advertising of tobacco products also seems a flawed approach to improving the health of the nation. Drugs are an undeniable scourge, but the biggest killers among them are tobacco (400,000 annually in the USA) and alcohol (90,000), not cocaine (8000) or heroin (6000), which nevertheless have more emotional impact and cause more alarm.

A further possibility for making risks more transparent is to list some activities that are estimated to increase one's chance of death by the same amount. The following table gives an example of the risks estimated to increase the chance of death in any one year by 1 in a million:

Activity	Cause of death
Smoking 1.4 cigarettes	Cancer, heart disease
Travelling 10 miles by bike	Accident
Flying 1000 miles by jet	Accident
One chest X-ray taken in a good hospital	Cancer
Eating 40lbs of peanut butter	Liver cancer caused by aflatoxin B
Drinking 30 12 oz cans of diet soda	Cancer caused by saccharin
Living 150 years within 20 miles of a nuclear power plant	Cancer

The following two formulations of a 'choice' problem given by two psychologists to practising physicians offer an interesting example of how subtle changes in the way which risks are expressed can have a major impact on perceptions and decisions:

(1) Imagine that the United States is preparing for the outbreak of an unusual Asian disease, which is expected to kill 600 people. Two alternative programs to combat the disease have been proposed. Assume that the consequences of the programs are as follows:

(a) If program A is adopted, 200 people will be saved.
(b) If program B is adopted, there is one-third probability that 600 people will be saved, and two-thirds probability that no people will be saved.

- Which of the two programs would you favour?

(2) Imagine that the United States is preparing for the outbreak of an unusual Asian disease, which is expected to kill 600 people. Two alternative programs to combat the disease have been proposed. Assume that the consequences of the programs are as follows:

(c) If program C is adopted, 400 people will die.
(d) If program D is adopted, there is one-third probability that nobody will die. And two-third probability that 600 people will die.

- Which of the two programs would you favour?

Most participants preferred program A over program B and preferred program D over program C, despite the formal equivalence of A and C and of B and D. The phrases *will be saved* and *will die* inspired very different reactions to the same problem.

Given a choice, people would rather not have to confront the gambles inherent in life's risky activities. Psychological research shows that whenever possible, people attempt to reduce the anxiety generated in the face of uncertainty by denying that uncertainty, thus making the risk seem either so small that it can safely be ignored or so large that it clearly should be avoided. They are uncomfortable when given statements of probability rather than fact; they want to know exactly what will happen. Tell people that their probability of developing cancer from a 70-year exposure to a carcinogen at ambient levels ranges between 0.00001 and 0.0000001, and their response is likely to be 'Yes, but will I get cancer if I drink the water?'

The way risks are presented also influences perceptions. You might be worried if you heard that occupational exposure at your job doubled your risk of serious disease compared to the risk working at some other occupation entailed; you might be less worried it you heard that your risk had increased from one in a million to two in a million. In the first case a *relative risk* is presented, in the second an *absolute risk*. Relative risk is generally used in medical studies investigating possible links between a risk factor and a disease—it is an extremely important index of the strength of the association between the factor and the disease, but it has no bearing on the probability that an individual will contract that disease. This may explain why airplane pilots who presumably have relative risks of being killed in airline crashes which are of the order of a thousand-fold greater that the rest of us occasional flyers, can still sleep easy in their beds. They know that their absolute risks of being the victim of a crash remains extremely small.

Many readers will be familiar with Thomas Jefferson's famous dictum to the effect that 'If we think (the people) not enlightened enough to exercise their control with a wholesome discretion, the remedy is not to take it from them, but to inform their discretion'. As in many Presidential utterances before and since, there is just the hint of the pompous and pious here, but the dictum certainly applies in the areas of risk assessment and risk appreciation. The general public's perceptions of risk are often highly inaccurate, and there is a clear need for educational programs. Those who plan such programs should remember that the general public's judgements of risk are sensitive to many influences and, for this reason, often differ markedly from expert's assessment of risk. In particular, perception of risk is greater for hazards whose adverse effects are uncontrollable, catastrophic, fatal (rather than simply likely to result in injury) and are not offset by compensating benefits. Furthermore, hazards associated with accidents in familiar and well-understood systems, for example, train crashes, may arouse less concern than those in an unfamiliar system (such as nuclear reactors), even if the former claim many lives and the latter result in few or no deaths. People perceive the nuclear reactor accident as a harbinger of further and possibly catastrophic mishaps. Ignoring such issues will mean that educating the public about risk will itself have a high risk of failure.

Unfortunately, many research studies have shown that people's beliefs change slowly and are extraordinarily persistent in the face of contrary evidence. Once formed, initial impressions tend to structure the way subsequent evidence is interpreted. New evidence appears reliable and informative if it is consistent with one's initial beliefs; contrary evidence is dismissed as unreliable, erroneous, or unrepresentative. (It is not, of course, only the general public who suffer from this malaise. The history of science is littered with examples of even the most distinguished scientists clinging desperately to their own pet theories in the face of overwhelming evidence that they are wrong.)

Perhaps the whole problem of the public's appreciation of risk evaluation and risk perception would diminish if someone could devise a simple scale of risk akin to the Beaufort scale for wind speed or the Richter scale for earthquakes. Such a 'riskometer' might prove very useful in providing a single number that would allow meaningful comparison among the risks from all types of hazards, whether they be risks due to purely voluntary activities (smoking and hang-gliding), risks incurred in pursuing voluntary but virtually necessary activities (travel by car or plane, eating meat), risks imposed by society (nuclear waste, overhead power lines, legal possession of guns), or risks due to acts of God (floods, hurricanes, lightning strikes). In another of his excellent and very readable books, *Innumeracy*, John Paulos proposes such a risk scale based on the number of people who die each year pursuing various activities. If one person in, say, N dies then the associated risk index is set at $log_{10} N$; 'high' values indicate hazards that are *not* of great danger, whereas 'low' values suggest activities to be avoided if possible. (A logarithmic scale is needed because the risks of different events and activities may differ by several orders of magnitude.) If everybody who took part in a particular pursuit or was subjected to a particular exposure died, then Paulos's suggested risk index would take the value zero, corresponding to certain death. In the UK, 1 in every 8000 deaths each year results from motor accidents; consequently the risk index for motoring is 3.90. Here are some examples of values for other events:

Event/activity	Risk index
Playing Russian roulette once a year	0.8
Smoking ten cigarettes per day	2.3
Being struck by lightning	6.3
Dying from a bee sting	6.8

As an indicator of things that should be regarded as dangerous we might use values below 3, whereas activities assigned values above 6 would be regarded as safe or of little concern in going about our daily lives.

Paulos's suggestion would clearly need refinement to make it widely acceptable and of practical use. Death, for example, need not be the only concern; injury and illness are also important consequences of exposure to hazards and would need to be quantified by any worthwhile index. But, if a satisfactory 'riskometer' could eventually be devised, it might reduce the chances of the public being mislead by media hype about lesser risks. And pronouncements about hazards and their associated risk index might become as familiar as statements that an earthquake that registered force 5 on the Richter scale has occurred or that winds of force 8 on the Beaufort scale are expected. The presentation of risks in this way in the media might largely overcome the criticism that journalists' current, often sensational approach, to risk presentation is largely responsible for the public's general misperception of risks. A riskometer would help improve media performance.

Acceptable Risk

Life is a risky business, and deciding which risks are worth taking and which should be avoided has important implications for an individual's lifestyle. The benefits gained from taking a risk need to be weighed against the possible disadvantages. Acceptable risk is proportional to the amount of benefits. Living life to the fullest means a balance between reasonable and unreasonable risks. In the end this balance is a matter of judgement and depends largely on the personality of the individual, so excessive dogmatism has little part to play. Nevertheless, it is very helpful to cultivate an ability to assess risks and compare them in a calm and rational manner.

For voluntary risks such as travelling across the country on a motor-cycle, we are probably reasonably adept at weighing the risk against the gain. In any case, voluntary risk is far more acceptable than involuntary—the public will accept about 1000-times-higher level of the former than of the latter. Of greater concern is assessment of the cost–benefit equation for risks to health caused by hazards over which we have little or no control. Many such hazards have recently featured prominently in scare stories in the media, although in most cases they have extremely small risks attached. Hurricanes, for example, cause about 90 deaths per year in the United States. If the average fatality corresponds to 35 years of lost life expectancy, the average American loses 0.5 day of life to the hurricane hazard. The corresponding figure for earthquakes is 0.1 for airline crashes is 1.0, and for chemical releases is 0.1.

Unreasonable public concern about a hazard, fuelled by the media, can cause governments to spend a good deal more to reduce risk in some areas—and a good deal less in others. Examples of where these disparities make little sense are not hard to find. In the late 1970s, for example, the United States introduced new standards on emissions from coke ovens in steel mills. The new rules limited emissions to a level of no more than 0.15mg/m^3 (milligram per cubic metre) of air. To comply with this regula-tion, the steel industry needed to spend $240 million a year. Supporters of the change estimated that it would save about 100 deaths from cancer and respiratory deaths a year, making the average cost per life saved $2.4 million. It is difficult to claim that this is money well spent when a large-scale program to detect lung cancer in its earliest stages, for example, might be expected to extend the lives of 7000 cancer victims an average of one year for $10,000 each and when the installation of special cardiac-care units in ambulances could prevent 24,000 premature deaths a year at an average cost of a little over $200 for each year of life. Assigning some economic weight to human lives is, of course, likely to be highly controversial, but it may be necessary if, in a system of finite resources, the aim is to achieve the most protection for the most people.

The media constantly draw health hazards to our attention, but constant preoccupation with hazards to health that have very small risks is unhealthy. People are in most ways safer today than ever before. In a very real sense, we can now indulge in the 'luxury' of worrying about subtle hazards that at one time, if they were even detected, would have been given only low priority beside the much greater hazards of the day. One hundred years ago, our great-great-grandparents, suffering the ravages of pneumonia, influenza and tuberculosis, would no doubt have been overjoyed to have replaced their health concerns with assessing the slight chance that calorie-cutting sweeteners in diet drinks may cause cancer.

According to Petr Skrabanek and James McCormick in their iconoclastic text *Follies and Fallacies in Medicine*,

> In any evaluation of risk it has to be remembered that life itself is a universally fatal sexually transmitted disease and that in the end nobody cheats death. A strong case can be made for living a life of modified hedonism so that we may enjoy to the full the only life which we are likely to have.

Such a view echoes that of the poet, William Drummond: 'Trust flattering life no more, reedeem time past, And live each day as if it were thy last'.

With this in mind I will now take a break from writing to enjoy a bottle of 1985, *Chateau Margaux*, consume a box of Belgium chocolates, make passionate love to my wife and afterwards smoke one or two of those 'herbal' cigarettes much loved in the sixties and seventies. (I will not, of course, inhale.)

Statistics, Statisticians and Medicine

12

To understand God's thoughts we must study statistics, for these are the measure of his purpose.
Florence Nightingale

Thou shalt not sit with statisticians nor commit a Social Science.
W.H. Auden

To be uncertain is uncomfortable, but to be certain is to be ridiculous.
Chinese proverb

Introduction

In my younger days, women at parties often used to ask what I did for a living. I proffered only the information that I worked somewhere called the Institute of Psychiatry, hoping, I confess, that they would jump to the conclusion that I was a psychologist or, better still, a psychiatrist. If they did, some question as to whether I was 'psychoanalysing them' often followed and this was usually a good predictor of a successful evening to come. But on the occasions where I was forced to reveal more about my job, until finally I had to admit to being a statistician, I usually left early to prevent the social isolation that almost always followed.

In the general public's affection, statisticians rank perhaps a trifle above politicians but certainly no higher than traffic police and tax collectors. Most people see statistics itself as dull, devious and even downright dangerous, and its practitioners are often judged professionally suspect and socially undesirable. Like most statisticians, I learned early in my career never to admit to other guests at parties what I do for a living.

One of the causes of the statistician's image problem is that most people have a distorted idea of what the great majority of statisticians actually do. For many the label *statistician* conjures up either a near-sighted

B. Everitt, *Chance Rules*, DOI: 10.1007/978-0-387-77415-2_12,
© Springer Science+Business Media, LLC 2008

character amassing volumes of figures about cricketers' bowling and batting averages, and about the number of home runs made in a baseball season or a government civil servant compiling massive tables of figures about unemployment, inflation, and Social Security payments, most of which are then ridiculed by less-than-well-informed journalists or radio and television commentators. But the majority of working statisticians (in the pharmaceutical industry, in medicine, in agriculture, in education etc.) are not compulsive 'number crunchers' but rather seekers of evidence and patterns among data subject to random variation. Statisticians are essentially professional sceptics who make decisions on the basis of probabilities (a bit like judges but with the advantage of knowing what probabilities are relevant.) Statisticians do not offer certainty, but by quantifying the uncertainty in any situation, they try to recognize a good bet. Sometimes, of course, they get it wrong! An excellent summary of what statistics is really about appears in the December 1998 issue of the *Journal of the American Statistical Association*, where Professor David Moore discusses its relationship to the liberal arts.

> Statistics is a general intellectual method that applies wherever data, variation, and chance appear. It is a fundamental method because data, variation and chance are omnipresent in modern life. It is an independent discipline with its own core ideas rather than, for example, a branch of mathematics........Statistics offers general, fundamental, and independent ways of thinking.

Statisticians have made, and are making, important contributions in many areas—statisticians really do count. But for our purposes here, it will be wise to concentrate on the impact they have had in a particular field. Medicine nicely fills the bill, because it is likely to be of relevance to most readers.

Medicine—from Anecdote to Evidence

> All who drink of this remedy recover in a short time, except those whom it does not help, who all die. Therefore, it is obvious that it fails only in incurable cases.

This aphorism is generally attributed to Galen (A.D. 130–200), a Greek physician, who was destined to dominate medicine for many centuries and who wrote with such conviction that few doctors dared criticize him. A prodigious writer, in one of his many books he gives an account of his own parents, describing his father as amiable, just and benevolent, and his mother as thoroughly objectionable, a woman who was always shouting at her husband and displaying her evil temper by biting her serving-maids. His father had a dream that his son was destined one day to become a great physician, and this encouraged him to send Galen to Pergamos and to Smyrna for a preliminary grounding in philosophy, and then onto Alexandria to specialize in medicine.

The aphorism indicates the kind of invulnerability claimed by physicians until well into the seventeenth century. Uncritical reliance on past experience, *post hoc, ergo propter hoc* ('after this, therefore because of it') reasoning, and veneration of 'truth' as proclaimed by authoritative people (particularly Galen) largely stifled any interest in experimentation or proper scientific exploration. Even the few who did attempt to increase their knowledge by close observation or simple experiment often interpreted their findings in the light of the currently accepted dogma. When, for example, Andreas Vesalius, a sixteenth century Belgian physician, first dissected a human heart and did not find the 'pores', that Galen said perforated the septum between the ventricular chambers, the Belgian assumed the openings were invisible to the eye. It was only several years after his initial investigation that Vesalius had the confidence to declare that the pores did not exist.

Similarly, the English physician William Harvey's announcement, in 1628, of the discovery of the circulation of the blood met with violent opposition; it contradicted Galen's view that blood flowed to and fro in a tide-like movement within arteries and veins. Even when it was rather grudgingly admitted that Harvey was probably right, a defender of the established view wrote that if the new findings did not agree with Galen, the discrepancy should be attributed to the fact that nature had changed. One should not admit that the master had been wrong.

Treatments Worthless—and Worse

For the medieval physician, choice of treatment depended largely on the results of observing one or two patients or on reports from colleagues, which also were based on very limited numbers of observations. But, because patients rather inconveniently vary in their responses to treatment, this reliance on a few specific cases led to the development of treatments that were often disastrously ineffective when applied more generally. Even the oath taken by Western physicians since the time of Hippocrates, in which they swear to protect their patients 'from whatever is deleterious and mischievous', did not stop many assaultive therapies from being administered or lessen the persistence of barbarous practices like copious bloodletting. Even the most powerful members of society were vulnerable to the ill-informed, if well-intentioned physician. At eight o'clock on Monday morning of 2 February 1685, for example, King Charles II of

England was being shaved in his bedroom. With a sudden cry he fell backward and had a violent convulsion. He lost consciousness, rallied once or twice, and after a few days, died. Doctor Scarburgh, one of the twelve or fourteen physicians called to treat the stricken king, recorded the efforts made to cure the patient:

> As the first step in treatment the king was bled to the extent of a pint from a vein in his right arm. Next his shoulder was cut into and the incised area was 'cupped' to suck out an additional eight ounces of blood. After this, the drugging began. An emetic and purgative were administered, and soon after a second purgative. This was followed by an enema containing antimony, sacred bitters, rock salt, mallow leaves, violets, beetroot, camomile flowers, fennel seed, linseed, cinnamon, cardamom seed, saphron, cochineal, and aloes. The enema was repeated in two hours and a purgative given. The king's head was shaved and a blister raised on his scalp. A sneezing powder of hellebore root was administered and also a powder of cowslip flowers 'to strengthen his brain'. The cathartics were repeated at frequent intervals and interspersed with a soothing drink composed of barley water, liquorice, and sweet almond. Likewise white wine, absinthe, and anise were given, as also were extracts of thistle leaves, mint, rue, and angelica. For external treatment a plaster of Burgundy pitch and pigeon dung was applied to the king's feet. The bleeding and purging continued, and to the medicaments were added melon seeds, manna, slippery elm, black cherry water, an extract of flowers of lime, lily of the valley, peony, lavender, and dissolved pearls. Later came gentian root, nutmeg, quinine and cloves. The king's condition did not improve, indeed it grew worse, and in the emergency forty drops of extract of human skull were administered to allay convulsions. A rallying dose of Raleigh's antidote was forced down the king's throat; this antidote contained an enormous number of herbs and animal extracts. Finally bezoar stone was given. 'Then', said Scarburgh, 'Alas! After an ill-fated night his serene majesty's strength seemed exhausted to such a degree that the whole assembly of physicians lost all hope and became despondent; still so as not to appear to fail in doing their duty in any detail, they brought into play the most active cordial'. As a sort of grand summary to this pharmaceutical debauch, a mixture of Raleigh's antidote, pearl julep, and ammonia was forced down the throat of the dying king.

Occasionally, serendipitous observations led to the discovery of more suitable treatments. An example is provided by the Renaissance surgeon, Ambroise Paré, while he was treating wounds suffered by soldiers during the battle to capture the castle of Villaine in 1537. Paré intended to apply the standard treatment of pouring boiled oil over the wound but ran out of oil. He then substituted a digestive made of egg yolks, oil of roses, and turpentine. The superiority of the new treatment became evident the day after the battle.

> I raised myself very early to visit them, when beyond my hope I found those to whom I applied the digestive medicament feeling but little pain, their wounds

neither swollen nor inflamed, and having slept through the night. The others to whom I had applied the boiling oil were feverish with much pain and swelling about their wounds. Then I determined never again to burn those so cruelly wounded by arquebusses.

The Beginnings of a Rational Approach

By the late seventeenth and early eighteenth centuries, scientists in general— and physicians in particular—began to adopt a more sceptical attitude to the pronouncements of authoritative figures, and medicine began to progress from dogmatic, even mystical, certainty to proper scientific uncertainty. One of the most notable example of the changes taking place during this period is provided by James Lind's investigations into the treatment of scurvy.

Scurvy is a disease characterized by debility, blood changes, spongy gums and haemorrhages in the tissues of the body. The symptoms come on gradually, beginning with failure of strength, and mental depression. Sallow complexion, sunken eyes, tender gums and muscular pains follow. These symptoms may persist, slowly worsening, for weeks. Teeth fall out and haemorrhages, often massive, penetrate muscles and other tissues. The last stages of scurvy are marked by profound exhaustion, fainting, and complications such as diarrhoea, pulmonary distress, and kidney troubles, any of which may bring about death. In 1932, it was discovered that scurvy is caused by deficiency of vitamin C and that even in desperate cases, recovery may be anticipated when the deficient vitamin is supplied, by injection or orally.

But 300 years ago physicians knew only that scurvy was common, was often fatal, and was a severe problem for mariners, causing more deaths in wartime than did the enemy. It is recorded, for example, that in 1740 Lord Anson took six ships on a world cruise and lost some 1200 of his men to the disease. There was some speculation that scurvy and diet were connected, but Lind was the first to investigate the relationship in a proper scientific fashion.

James Lind was a Scottish physician who took his M.D. degree at Edinburgh in 1748 and was physician at the Haslar hospital for men of the Royal Navy, Gosport, Hampshire, England, from 1758 until his death. In his book, *A Treatise on the Scurvy*, published in 1754, he gives the following description of his landmark study.

On the 20th May 1747, I took twelve patients in the scurvy, on board the Salisbury at sea. Their cases were as similar as I could have them. They all in general had putrid gums, the spots and lassitude, with weakness of their knees. They lay together in one place, being a proper apartment for the sick in the fore-hold; and had one diet in common to all, viz. water-gruel sweetened with sugar in the morning; fresh mutton broth often times for dinner; at other times puddings, boiled biscuit with sugar etc. And for supper, barley and raisins, rice and currants, sago and wine, or the like. Two of these were ordered each a quart of cider a day. Two others took twenty-five gutts of elixir vitriol (*sulfuric acid*) three times a day, upon an empty stomach; using a gargle strongly acidulated with it for their mouths. Two others took two spoonfuls of vinegar three times a day, upon an empty stomach: having their gruels and their other food well acidulated with it, as also the gargle for their mouths. Two of the worst patients, with the tendons in the ham rigid (a symptom none of the rest had) were put under a course of sea-water. Of this they drank half a pint every day, and sometimes more or less as it operated, by way of a gentle physic. Two others had each two oranges and one lemon given them every day. These they eat with greediness, at different times, upon an empty stomach. They continued but six days under this course, having consumed the quantity that could be spared. The two remaining patients, took the bigness of a nutmeg three times a day of an electuary recommended by a hospital-surgeon, made of garlic, mustard-feed, rad. raphan, balsam of Peru, and gum myrr; using for common drink barley water well acidulated with tamarinds; by a decoction of which, with the addition of cremor tartar, they were greatly purged three or four times during the course. The consequence was, that the most sudden and visible good effects were perceived from the use of the oranges and lemons; one of those who had taken them, being at the end of six days fit for duty. The spots were not indeed at that time quite off his body, nor his gums sound; but without any other medicine, than a gargle of elixir vitriol, he became quite healthy before we came into Plymouth, which was on the 16th June. The other was the best recovered of any in his condition; and being now deemed pretty well, was appointed nurse to the rest of the sick.

In spite of the relative clear-cut nature of his findings, Lind still advised that the best treatment for scurvy involved placing stricken patients in 'pure dry air.' No doubt the reluctance to accept oranges and lemons as treatment for the disease had something to do with their cost compared to the 'dry air' treatment. In fact, it was a further 40 years before Gilbert Blane, Commissioner of the Board of the Care of Sick and Wounded Seamen, succeeded in persuading the Admiralty to make the use of lemon juice compulsory in the British Navy. But once again, the question of cost quickly became an issue, and limes, which were cheaper, were substituted for lemons. Economy thus condemned British sailors to be referred to, for the next 200 years as 'limeys'.

What makes Lind's investigation so notable for the time is its comparison of different treatments and the similarity of the patients at the commencement of the study—that is, they were all at a similar stage of the illness and were all on a similar diet. As we shall see later, such characteristics, are much like those demanded in a modern clinical trial.

A further example of a more rational approach to medical questions is provided by the work of Florence Nightingale. Born on 12 May 1820 in Florence, Italy, but raised in England, Florence Nightingale trained as a nurse in Kaiserworth and Paris. In the Crimean War, after the battle of Alma (1854), she led a party of thirty-eight nurses to organize a nursing department in Istanbul. There she found grossly inadequate sanitation, but she soon established better conditions and had 10,000 wounded under her care. In 1856, she returned to England, where she established the Nightingale Training School for nurses at St. Thomas' Hospital, London and spent several years on army sanitary reform, the improvement of nursing, and public health in India.

Florence Nightingale

In her efforts to improve the squalid hospital conditions in Turkey during the Crimean War, and in her subsequent campaigns to reform the health and living conditions of the British Army, the sanitary conditions and administration of hospitals and the nursing profession, Florence Nightingale was not unlike many other Victorian reformers. But in one important respect she was very different: She marshalled massive amounts of data, carefully arranged and tabulated, to convince ministers, viceroys, secretaries, undersecretaries and parliamentary commissioners of the truth of her cause. No major national cause had previously been championed by the presentation of sound statistical data. One telling example will serve to illustrate Florence Nightingale's approach: she found that the British soldiers who succumbed to unsanitary living conditions before Sebastopol were about seven times the number felled by the enemy. Those who opposed her reforms went down to defeat because her data and statistics were unanswerable and their publication led to an outcry. (One of the diagrams used for presenting statistical material is reproduced below)

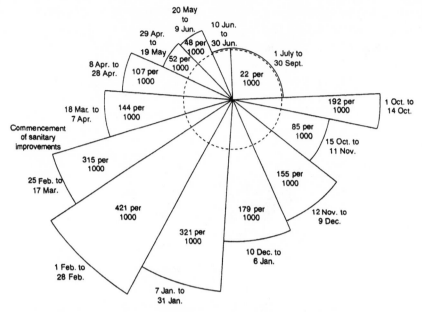

A diagram representing mortality in the hospitals at Scutari and Kulali from October 1, 1854 to September 30, 1855. The area within the dashed circumference represents the average mortality rate in the military hospitals in London, 20.9 per 1000

Florence Nightingale was instrumental in the founding of a statistical department in the army. Her pioneering use of statistical tables and charts to back her call for reform places her firmly in the forefront of the history of statistics. Although more widely known as the Lady of the Lamp, she is also recognized as the Passionate Statistician. In 1907, she became the first woman to be awarded the Order of Merit. Three years later she died peacefully at the age of 90.

The early twentieth century has seen most clinicians come to appreciate a systematic and formal approach to the evaluation of different treatments. They have realized that individual differences in the reactions of patients to diseases and therapies are often extremely variable and that personal observations on a small number of patients, however acutely and accurately made, are unlikely to tell the whole story.

In the past, the 'personal observation' approach has led to the establishment of such 'treatments' as bloodletting, purging and starvation. But, even more recently, the results of relatively personal observations on extended sequences of patients in inappropriately designed studies have lead physicians astray. An example from psychiatry will illustrate what can go wrong and the sometimes tragic results for patients.

In the 1920s, Dr. Henry A. Cotton proposed the theory of focal infection to explain mental disorders. His proposal was based on observation of many psychiatric patients. According to Dr. Cotton,

> The so called functional psychoses we believe today to be due to a combination of many factors, but the most constant one is the intra-cerebral, bio-chemical cellular disturbance arising from circulating toxins originating in chronic foci of infection, situated anywhere in the body, associated probably with secondary disturbance of the endocrin system. Instead of considering the psychosis as a disease entity, it should be considered as a symptom, and often a terminal symptom of a long continued masked infection, the toxaemia of which acts directly on the brain.

Dr. Cotton identified infection of the teeth and, tonsils are the most important foci to be considered, but the stomach and, in female patients, the cervix could also be sources of infection responsible, according to Dr. Cotton's theory for the mental condition of the patient. The logical treatment for the mentally ill resulting from Dr. Cotton's theory was surgical elimination of the chronically infected tissue, all infected teeth and tonsils certainly, and for many patients, colectomies. Additionally, female patients might require enucleation of the cervix, or, in some cases, complete removal of fallopian tubes and ovaries. Such treatment, although extremely severe was, according to Dr. Cotton, enormously successful, with out of 1400 patients treated only 42 needing to remain in hospital.

The focal infection theory of functional psychoses was not universally accepted, neither were the striking results said by Dr. Cotton to have been obtained by the surgical removal of these infections. Several later studies cast grave doubts over removal of focal infections as a treatment for some types of mental illness, and eventually drove a nail into the coffin of Dr. Cotton's theory as to the cause of these conditions. The drastic, even cruel treatment, proposed by Dr. Cotton for the mentally ill has long since been abandoned.

Clearly then, there is a need for some form of adequately controlled procedure by which to assess the advantages and disadvantages of different treatments. This need has become even more acute over the last two decades, with the steadily increasing production of new and powerful drugs designed to relieve suffering in many areas. The need has largely been met by the use of the controlled clinical trial.

The Clinical Trial

If a doctor claims that a certain type of psychotherapy will cure patients of their depression, or that vitamin C prevents or cures the common cold, how should these claims be assessed? How do we know whether the drug AZT is likely to prevent the development of AIDS, or whether radiation therapy will prevent the recurrence of breast cancer? What sort of evidence do we need to decide that claims made for the efficacy of clinical treatments are, indeed, valid? One thing is certain: We should not rely either on the views of 'experts' unless they produce sound empirical evidence (measurements, observations—*data*) to support their views, nor should we credit the anecdotal evidence of people who have had the treatment and, in some cases, been 'miraculously' cured (and it should be remembered that the plural of anecdote is not evidence). Such 'wonder' treatments (which are often exposed as ineffectual when subjected to more rigorous examination) are particularly prevalent for those complaints for which conventional medicine has little to offer, as we shall see in the next chapter.

There is clearly a need for some form of carefully controlled procedure for determining the relative effects of differing treatments. Such a need has largely been met by the development of the clinical trial, which is a medical experiment designed to evaluate which (if any) of two or more treatments is the more effective. The three main components of a clinical trial are:

(1) Comparison of a group of patients given the treatment under investigation (the treatment group) with a control group of patients given either an older or standard treatment if one exists, or an 'inert treatment' generally known as a *placebo*. (Simply observing the group of patients who receive the new treatment would not exclude spontaneous recovery as a possible reason for any improvement that occurs.)

(2) An acceptable method of assigning patients to the treatment and control groups.

(3) A means of assessing effectiveness, such as disappearances of symptoms, quicker recovery, lower number of deaths, and the like. (In general, some numerical measure will be needed.)

A crucial aspect of the design of a clinical trial is deciding how patients should be allocated to the competing treatment groups in a trial. The objective is that such groups should be alike in all respects except the treatment received. Should the clinician decide which patient goes into which group? Possibly, but then the results of the trial would be viewed with a considerable amount of scepticism. The clinician, might, for example, allocate the patients with the worst prognosis to the (in his or her opinion) promising new therapy and the more mildly affected patients to the older treatment. Or older patients might receive the traditional therapy and youngsters the new one, and so on. All of these procedures would tend to invalidate any findings. Should the patients themselves decide what treatment to receive? Again, this would be highly undesirable. Patients are likely to believe that the new therapy might solve all of their problems. Why else would it be featured in the trial? Thus few patients would knowingly select a placebo.

Should the first patients to volunteer to take part in the trial should all be given the novel treatment, and the later ones used as controls? Again, early volunteers might be more seriously ill— desperate to find a new remedy that works. What about simply alternating patients, putting every other one into the control group? The objection to this is that the clinician will know who is receiving what treatment and may be tempted to 'tinker' with the scheme to ensure that his patients who are most ill receive the new treatment.

In a modern controlled trial, none of the ad hoc allocation procedures considered above (or others that might be thought of) would be regarded as satisfactory. Instead, the treatment to be given to each individual patient is decided by chance. Allocation to treatment group or control group is random. It *could* be arranged by flipping a coin each time a patient enters the trial and allocating the patient to the new treatment if the result is a head or to the control group if a tail appears. In practice, a more sophisticated randomization procedure is generally used. The essential feature is the randomization rather that the mechanism used to achieve it. It was Sir Ronald Aylmer Fisher, arguably the most influential statistician of the twentieth century, who introduced the principle of random allocation.

Born in East Finchley, London, on 17 February 1890, Fisher won a scholarship in mathematics to Cambridge in 1909. He made great contributions both to the theory of statistics and to genetics, but perhaps his most important contribution was the introduction of the randomization principle. Fisher's early applications of randomization were in agriculture. To determine which fertilizers effected the greatest crop yields, Fisher divided agricultural areas into plots and randomly assigned the plots to different experimental fertilizers.

Before the introduction of randomization, most of the early experiments to compare competing treatments for the same condition (Lind's study, for example) involved arbitrary, nonsystematic schemes for assigning patients to treatments. The first trial wherein a properly randomized control group was employed was a clinical trial assessing the efficacy of streptomycin in the treatment of pulmonary tuberculosis; it was carried out in the UK in the late 1940s.

Randomization serves several purposes:

(1) It provides an impartial method, free of personal bias, for the assignment of patients to treatment groups. This means that treatment comparisons will not be invalidated by the way the clinician might choose to allocate treatments if left to use his or her own judgement.
(2) It tends to balance treatment groups in terms of extraneous factors that might influence the outcome of treatment—even in terms of those factors the investigator may be unaware of.

photo A. Barrington-Brown (c) R.A. Fisher Memorial Trust.

Sir Ronald Aylmer Fisher

(Randomization is an excellent principle but one that cannot always over-come other factors in a study. A classic example of what can go wrong occurred in the Lanarkshire milk experiment of the 1920s. In this trial, 10,000 children were given free milk supplementation, and a similar number received no supplementation. The two groups were formed by random allocation. Unfortunately, however, well-intentioned teachers gave the poorest children free milk rather than sticking strictly to the original groups. As a consequence, the effects of milk supplementation could not be distinguished from the effects of poverty.)

The randomized clinical trial becomes even more appealing when the participants in the trial do not know which treatment they are receiving. This removes a further potential source of bias. When both patient and doctor are unaware of the treatment, the trial is said to be *double-blind*.

The double blind, randomized, controlled clinical trial is now the 'gold standard' for evaluating the effectiveness of treatments. Let's look at two examples (for one of which we have to interpret 'treatment' in the widest sense).

Vitamin C and the Common Cold

In the later stages of his life, Linus Pauling, twice winner of the Nobel Prize, championed the use of vitamin C for preventing a variety of unpleasant conditions ranging from the common cold to cancer. In his book *vitamin C and the common cold*, for example, he claimed that ascorbic acid (vitamin C), in large daily doses could prevent upper respiratory infections (the common cold). As a prophylactic measure, Pauling recommended taking 1–3 g daily. His advice was largely based on a combination of personal experience and evolutionary considerations. But even the views of a scientist as eminent as Professor Pauling are not acceptable unless backed up by the results from a suitable clinical trial. Several trials of the effectiveness of vitamin C for the common cold have been carried out; the account that follows is based on the trial reported in 1975 by Dr. Lewis and his colleagues. In addition to the results actually obtained, some additional hypothetical 'results' will be presented to illustrate a number of important points.

In the Lewis study, 311 people were randomly assigned to four groups. Group One received a placebo both for maintenance during the time they did not have a cold and for therapy when they contracted a cold. Group Two received placebo for maintenance but ascorbic acid for colds. Group Three received ascorbic acid for maintenance and placebo for colds, and Group Four received ascorbic acid for both maintenance and therapy. The study continued for 9 months. The following measures of effectiveness were used: number of colds per person, mean duration of colds and time spent at home suffering from colds.

Before presenting and examining the results actually obtained from the study let's pretend that the following results were collected from a subset of the patients in the study:

	Group Two	Group Four
Number of subjects having at least one cold in 9-month period	20	10
Number of subjects in group	50	50

Because double the number of patients in Group Two as in Group One suffer at least one cold during the 9 months of the trial, you might jump to the conclusion that receiving vitamin C both for maintenance therapy is clearly preferable–you might even be right! But imagine the study being repeated with a different set of subjects. Subjects vary in their responses to treatment—that is, they demonstrate the random variation described in several previous chapters—so it is possible that the results might now look like this:

	Group Two	Group Four
Number of subjects having at least one cold in 9-month period	15	10
Number of subjects in group	50	50

Group Four still outperforms Group Two, but the difference is now not so clear. The job of the statistician is to assess whether the *observed* difference on the particular set of patients used in the trial reflects a *true* difference amongst all similar subjects who may have taken part in the trial or is simply attributable to chance when there is *no* true difference. The result found for the particular set of subjects used (which is called the sample) is not in itself of great interest except in what it says about the difference among *all* the subjects who might have taken part (called the population).

The statistician's approach is first to assume that in the population, there is *no* difference between the treatments. The probability of seeing the *observed* difference under this assumption is then calculated (how this is done need not concern us here). If this probability is small, the statistician will conclude that the 'no difference' assumption under which it was calculated is probably false; otherwise the assumption will be accepted as reasonable. If the former is the case, the statistician declares that a treatment difference has been demonstrated in the trial; if the latter, the inference is that the treatments do not differ in effectiveness. The only remaining question is 'what constitutes a small probability?' Conventionally, this is set at the value 0.05 but this value is not cast in stone and may need to be changed, depending on the detailed needs of a particular study.

Note that the statistician does not claim certainty in deciding for or against a difference in treatment effectiveness. Random variation in subjects' responses makes certainty impossible. The procedure outlined above does, however, enable the statistician and the clinician involved with the trial to avoid making overly optimistic claims about the effectiveness of new treatments, while retaining a good chance of identifying those treatments that do produce better results. The formalized approach adopted (it is known as significance testing) gives the best chance of not being overly optimistic *or* pessimistic about the possible advantages of a new treatment.

Now let us return to the *actual* results obtained in the Lewis trial of vitamin C. The table that follows gives the average duration of colds in each of the four treatment groups. The cold duration averaged 7.1 days for the subjects in the placebo group, 6.6 days for those taking 3 g of vitamin C per day either as maintenance (prophylactic) or therapy (therapeutic), and 5.9 days taking vitamin C for both therapy and maintenance. After applying the appropriate significance-testing procedure, the authors concluded that there was some evidence that vitamin C treatment might shorten the duration of colds to a small degree.

Group	Prophylactic	Therapeutic	Number of patients	Average duration of cold (days)
One	Placebo	Placebo	65	7.1
Two	Placebo	Vitamin C	56	6.5
Three	Vitamin C	Placebo	52	6.7
Four	Vitamin C	Vitamin C	76	5.9

Healing Power of Prayer?

Prayer can range from a personal thanksgiving addressed to God or a god for an earnest request or devout wish. Prayer is clearly, for most believers, a very personal pastime involving prayers about themselves and their family; there are occasions, however, on which the faithful are called on to pray for others. Richard Dawkins in his book *The God Delusion* points out that prayers are frequently offered both privately and in formal places of worship, for sick people unknown to the person who prays for them. Dawkins then recounts the story of Darwin's cousin, Francis Galton, addressing the question as to the efficacy of such prayers. Galton noted that every Sunday, in churches throughout Britain, entire congregations prayed publicly for the health of the royal family and argued that the members of this family should therefore be unusually fit compared to the rest of us who are only usually prayed for by our nearest and dearest. According to Dawkins, Galton actually looked into this possibility but found no evidence of a difference. One suspects that if Galton really did do this his tongue was very firmly in his cheek.

But over the last decade or so, several investigators have attempted to put the question of whether praying for sick people improves their health, to rigorous scientific scrutiny via the double–blind, randomized clinical trial. Here we shall concentrate on one such study reported in 2001 in that august publication, the *British Medical Journal* (BMJ). The title of the paper is 'Effects of remote, retroactive intercessory prayer on outcomes in patients with bloodstream infection: randomised controlled trial'. All adult patients whose bloodstream infection was detected at a university hospital during 1990–1996 were included in the study. Bloodstream infection was defined as a positive blood culture (not resulting from contamination) in the presence of sepsis. In July 2000 all the records of patients from 4 to 10 years earlier were randomized into two groups and the intervention group decided by the toss of a coin. A list of the first names of the patients in the intervention group was given to a person who said a short prayer for the well-being and full recovery of the group as a whole.

The results from the study showed that there was no statistically significant difference in mortality in the two groups but that average length of stay in hospital and average duration of fever were significantly shorter in the group of patients who had been prayed for. The author of the paper concluded:

> No mechanism known today can account for the effects of remote, retroactive intercessory prayer said for a group of patients with bloodstream infection. However the significant results and the flawless design prove that an effect was achieved Remote, retroactive intercessory prayer can improve outcomes in patients with bloodstream infection. The intervention is cost effective, probably has no adverse effects, and should be considered for clinical practice.

Consider for a moment what is being suggested here, namely that prayer carried out 4–10 years after the patients' infection and hospitalisation can have a retrospective effect on their health; The implications of the author's conclusions for science, if true, would be monumental. The implication would still be enormous if we altered the interpretation of the results to suggesting that prayer has determined the toss of the coin used to ensure that the intervention group contains what were, 4–10 years ago, the healthier patients. It will come as no surprise that the paper generated a lot of correspondence; some samples follow:

• According to Clause 30 of the latest revision of the Declaration of Helsinki: At the conclusion of the study, every patient entered into the study should be assured of access to the best proven prophylactic, diagnostic and therapeutic methods identified by the study. To meet this ethical standard, the prayer should now be said for the control group. If the treatment is effective this should have the effect of removing the difference between the groups. I await the results with interest.

• Congratulations-this is a first, Evidence of Providence Based Medicine

• Fascinating article 'So explain to me how this thing works. . . . . .are you telling me that if I pray for my dead Aunt Mildred perhaps I can get her to retroactively change her will to include me this time?

- The author deserves notice for a remarkable contribution, if not to science, to ethics or to consistency, perhaps to humour, of a sort.

- It may well be that the efficacy of prayer was underestimated– perhaps greatly underestimated–in this report. As spelled out in the Methods section, 'A list of the first names of the patients in the intervention group was given to a person who said a short prayer for the well being and full recovery of the group as a whole.' Now, it is highly likely that some-- perhaps many–of the control group members shared some of those first names. Thus an undetermined number of control group members may have inadvertently benefitted from the intercessory prayer, thereby spuriously bringing the control group's data into closer agreement with the intervention group's. Indeed, although other commentators here have decried the researcher's possibly unethical behavior in with- holding effective treatment (prayer) from the control group, this trial may in fact already have had vast, unmeasured, beneficial effects on a large segment of the general public: those who share first names with the intervention group. One could ascertain whether such collateral effects have actually occurred, by comparing hospital records for patients whose first names do or do not appear on the list. Such a tally of several hundred thousand records from many hospitals would, I venture to predict, reveal statistically significant differences (albeit perhaps for conditions other than bloodstream infection and outcome measures other than duration of hospital stay).

- So let me get this right. . .you want to allocate £10 K in the budget for an end-of-year prayer session for all patients who have been through ITU? Yes And this will reduce out ITU stay by one day per patient prayed for? Yep And if were over budget by the end of the year, and cant afford the prayer session it doesnt matter as well still get the therapeutic effect-just so long as we make a solemn promise to pray for everyone the year after instead? Yep But we could put it back a year or two if were still short of cash.. Yes-just so long as we squeeze in a prayer session sometime before 2013 we should get the benefits.

I will only make this observation about the paper-it appeared in the BMJ in December but should perhaps have appeared in early April.

There have of course been other investigations into the healing power of prayer that cannot perhaps be dismissed as rapidly as the one just described. In most cases, the results have not favoured the group of patients prayed for although a number *have* showed a small but statistically sig- nificant effect in favour of these patients. But to change my own subjective prior probability that intercessory prayer can aid healing from its current very, very low value to something closer to 1 would require indisputable evidence of *large* effects from *many* well-designed and impeccably performed studies. Currently this is not happening and I remain extremely sceptical that it ever will.

Summary

The introduction of Fisher's idea of randomization into the comparison of different treatments was a breakthrough, but 50 years and many, many randomized trials later, there are still clinicians—and prospective participants in such trials—who have concerns about the procedure. The reasons for these concerns are not difficult to identify. The clinician who is responsible for restoring the patient to health, and who suspects that any new treatment is likely to have advantages over the old, may be unhappy that many patients will be receiving what she views as the less valuable treatment. The patient who is recruited for a trial, having been made aware of the randomization component, may worry about the possibility of receiving an 'inferior' treatment. The clinician's ethical dilemma is understandable but misplaced. Doctors who are troubled by randomized trials might do well to take a wider view and accept the uncertainty involved in *much* of what they practise. Most would be forced to admit to themselves and their patients that, in many circumstances, they are unsure what is the best action to take. It is a sad and ironic fact that even with the advent of clinical governance, the National Institute for Clinical Excellence (NICE) in the United Kingdom and the like, it remains the case that a clinician can promote a vast range of therapies to his or her patients, and is rarely called on to demonstrate the effectiveness or efficacy of what he or she does. If challenged, the old icon of 'clinical freedom' can be invoked, or, if all fails, past experience. But past experience can be misleading.

As Sir Iain Chalmers has memorably pointed out, if one decides to give all one's patients a particular treatment, there are few, if any, people around who will counsel caution, or even be in a position to stop you. But woe betide the clinician who admits uncertainty, and so wishes to undertake a clinical trial. 'If I give all my patients the same treatment, no one is around to stop me, but should I decide to give only half of my patients the very same treatment, the world seems full of people who will tell me why I should not do this'. It seems that a clinician needs permission to give a new drug to half his or her patients, but not to give it to all of them!

When doctors *are* able to admit to uncertainty in many of their practices, then no conflict exists between the roles of the doctor as clinician and doctor as scientist. In such circumstances, it cannot be less ethical to choose a treatment by random allocation within a controlled, trial than to be guided by what happens to be readily available, by hunch or what a drug company recommends. The most effective argument in favour of randomized clinical trials is that the alternative, practising in complacent uncertainty, is worse.

The randomized, controlled, clinical trial is perhaps the greatest contribution statisticians have made to twentieth-century medical research. It is the classic example of what has recently become almost an evangelical movement amongst many clinicians—evidence-based medicine. Nowadays about 8000 such trials are performed each year. But there is one area that has remained peculiarly resistant to subjecting itself to the rigorous examination that clinical trials afford—this is the area of alternative or complementary medicine. The possible reasons behind this resistance are taken up in the next chapter.

(Chapter 13 is somewhat of a digression from the main theme of this book. It is an attempt to expose the general lack of statistical respectability found in the evaluation of alternative therapies, as an example of the many areas where anecdote, dogma and even outright prejudice hold sway over reasonable scepticism and scientific curiosity. Unless they are willing to subject their beliefs to scrutiny, those readers convinced that regular doses of reflexology keep them in perfect health, are advised to move speedily on to Chapter 14.)

Alternative Therapies—Panaceas or Placebos?

Formerly, when religion was strong and science weak, men mistook magic for medicine; now when science is strong and religion weak, men mistake medicine for magic.
Thomas Szasz

In California everyone goes to a therapist, is a therapist, or is a therapist going to a therapist.
Truman Capote

Fortunately, analysis is not the only way to resolve inner conflicts. Life itself remains a very effective therapist.
Karen Horney

Introduction

Magic and medicine have always been closely linked. The African witch doctor, the Native American Indian medicine man and Europe's medieval alchemists—all were a mixture of magician and physician. Any success they achieved in the treatment of their patients probably resulted from observing the effects of often nauseating remedies on a small number of these patients. Patients may have recovered simply to avoid being subjected to a further dose of some revolting compound! But in many cases they probably recovered either because their disease ran its natural course or because of what has become known as the placebo effect. This term refers to some benefit, following from treatment, that arises when the physician's belief in the treatment and the patient's faith in the physician exerting a mutually reinforcing effect, rather than from any direct result of the treatment itself. In the nineteenth century, for example, a lady brought her doctor to court for presenting her with a bill for administering morphine for a particular condition, when later she discovered he had, in fact, injected only water. Although the lady's condition had nevertheless been cured, the court found against the doctor. (Today this story is an example of what has become to be known as 'Asher's Paradox', after the distinguished London physician, Richard Asher, who claimed that 'if you can believe fervently in the treatment you are suggesting to a patient, even though controlled studies show that it is quite useless, then your results are much better, your patients are much better, and your income is much better too'. Asher goes on to suggest that this accounts for the remarkable success of some of the less gifted, but more credulous members of the medical profession, and also for the often violent dislike of statistics and controlled tests which fashionable and successful doctors are accustomed to display.)

The history of medicine is littered with claims for wonderful results from extraordinary treatments, most of which almost certainly arise from the mistaken assumption that an alteration in symptoms after the administration of a treatment is necessarily a specific result of that therapy—the placebo effect in action. One of my favourites is described by the late Petr Skrabanek and James McCormick in their wonderful book Follies and Fallacies in Medicine.

> In the twenties, Professor Eugene Steinach of Vienna introduced vasectomy as a rejuvenating procedure, his rationale being that as loss of sperm had a debilitating effect (a popular belief), it would surely be the case that blockage of the loss would have invigorating results. As a result of the 'success' of this operation over one hundred teachers and university professors underwent the procedure. Their number included Sigmund Freud and the poet W.B.Yeats.

Many readers may indulge themselves in a smug smile at the absurdity of even the 'great and the good' 70 years ago, confident that in the early twenty-first century, the days of quacks relying on a gullible public and the placebo effect to peddle their wares, have all but disappeared. Such readers should ponder the following list:

(1) In 1981, President Ronald Reagan was amongst those who congratulated Oral Roberts, the faith healer, on the foundation of his university, the City of Faith.

(2) In the UK in 1982, His Royal Highness Prince Charles became President of the British Medical Association and exhorted the profession to return to the precepts of *Paracelsus* whose pharmacopoeia included such cures as *zebethum occidentale*—dried human excrement.

(3) In 1995, the Dean of the Faculty of Homoeopathy in Great Britain prescribed kitchen salt so diluted that there is unlikely to be a molecule of sodium chloride in a large barrel, to help 'a girl with a broken love affair or a woman who has never been able to cry*cdots* to unwind.'

(4) In 1998, Mr. Glen Hoddle, one–time coach/manager of the English football (soccer) team and an expert on reincarnation, employed the services of a faith healer, Eileen Drewery, to aid injured players.

(5) In 1997, a judge in Maglie, a small town in Southern Italy, decreed that the country's health service must pay for cancer patients to take an expensive, untested remedy promoted by a retired physiologist named Luigi Di Bella. The treatment's unanimous rejection by scientists, the judge ruled, was not a good enough reason to deprive patients of it.

In the United States alone, at least 10 billion is spent annually on what are usually referred to as alternative therapies. Skrabanek and McCormick loosely classify such treatments into five categories:

(1) Mind cure: all forms of faith healing, Christian Science, etc.
(2) Medication: homoeopathy, herbalism, etc.
(3) Manipulation: osteopathy, chiropractic, reflexology, acupuncture, etc.
(4) Occultism: psychic surgery, medical dowsing, etc.
(5) Quack devices: negative ionizers, ozone generators, electroacupuncture devices, etc.

Several of these (for example, acupuncture, homoeopathy, osteopathy and chiropractic) are now used so commonly that many readers may be surprised to see them lumped with psychic surgery and medical dowsing, clear examples of 'quackery' to most people. Many readers may indeed be convinced that they have been helped by one or more of these treatments. And even the author has been known to resort to two of them when anxious to overcome, as quickly as possible, back problems and minor injuries caused by too much running. Conventional medicine has little to offer people suffering from lower-back pain, or the pain of arthritis and, not surprisingly, patients with these problems are likely to be the greatest users of alternative therapies. Desperate people will try desperate measures. And they may even write to the papers to tell others about their miraculous cure; the following, for example, is a letter published in 2007 in The Observer, a UK Sunday paper:

> Orthodox medicine watched as my asthma of some 10 years standing worsened every year with more serious and frequent attacks, despite their prescriptions. I turned to homoeopathy and my asthma was cured-and that is not a word I use lightly. Attacks diminished immediately and I have had none in 12 or 13 years. The power of homoeopathy is formidable. Only people who know nothing about it can say it has no scientific value-it is more in tune with Einstein than Newton

But anecdotes eulogizing such procedures and therapies are no substitute for hard evidence of the kind that would be produced by randomized clinical trials. Anecdotes are uniquely uninformative—even dangerous— when it comes to generalizable matters of health. And if we accept them in support of complementary medicine and reject them for orthodox medicine, we are de facto introducing a double standard into the evaluation of health care.

Unfortunately, it is only quite recently that clinical trials of alternative therapies have been performed, and many practitioners of alternative medicine have taken exception to even the relatively few trials that have been conducted. The following objections are typical:

(1) Experience has shown that should there be scepticism and doubt in the mind of a third party close to the patient........failure is usually inevitable.

(2) Due to different belief systems and divergent theories about the nature of health and illness, complementary and alternative medicine disciplines have fundamental difference in how they define target conditions, causes of disease, interventions and outcome measures of effectiveness.

Both statements seem to be attempts to pre-empt the suggestion that the remedies of alternative medicine should, like other treatments, be subjected to proper scientific investigation by applying the clinical trial methodology. The implication is that such methodology is not appropriate in this case and that even consistently negative findings will not undermine the alternative therapists' faith in their remedies. In response, a quotation from Anthony Garrett, printed in the journal The Skeptic, might be appropriate: 'In a society as open and susceptible to fraud as ours is, the truth needs all the help it can get'.

(Supporters of conventional medicine are themselves not entirely blameless for the relative dearth of acceptable studies of alternative procedures. They have often failed to allocate the necessary financial resources to undertake the relevant studies, perhaps because of their heavy investment in defending the status quo against the unconventional.)

But even if the advocates of alternative medicine (and some dyed-in-the-wool conventional clinicians) appear to be uninterested in the scientific evaluation of alternative treatments, the rest of us might like to look at any hard evidence, either for or against them. Here we shall consider only acupuncture and homoeopathy trials.

Acupuncture

Acupuncture is a type of surgical procedure devised in China many centuries before Christ. It consists of the insertion of needles of various metals and shapes into one or several of 365 spots, corresponding to the days of the year, that are specified for this purpose on the human trunk, the extremities and the head. Over the years it has been used for an impressive variety of diseases, in particular for the treatment of arthritis, headache, convulsions, lethargy, and colic. In its early manifestation (the third to the first centuries B.C.), acupuncture was used as a form of bloodletting in a magico-religious ritual during which the malevolent spirit of disease was allowed to escape. Later, the concept of vital energy (pneuma, Qi) was introduced and was said to flow in channels beneath the surface of the body. The surface markings of these channels are known as meridians. The acupuncture points (acupoints) are located along these meridians in sites where the Qi channels can be directly tapped by the needle. The stimulation of the acupoints is considered to be able to not only release an excess of Qi but also to correct a deficiency, thus maintaining harmony between the opposing metaphysical principles of Yin and Yang.

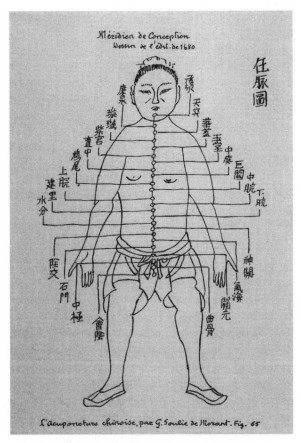

Ancient diagram of acupuncture sites

The fact that acupuncture has been used for over 5000 years, particularly in China, is one of the points often used in support of its effectiveness, the argument being that if it's been around so long there must be something to it. (If this is going to be our standard of judgement, we may as well be similarly convinced by astrologers who claim that Saturn being on the cusp of Mercury at the time of your birth preordains that you shall go through life as, perhaps, a statistician.) In fact acupuncture in China was rejected by the Emperor in 1822 as a bar to the progress of medicine and was removed from the curriculum of the Imperial Medical College. Its wholesale reintroduction took place in the 1960s and 1970s, as a political ploy by Maoists to provide cheap medical care for millions in newly 'liberated' areas. In the West, acupuncture has been accepted and then rejected again at least four times in the last 300 years. Its current wave of acceptance and popularity began in 1972 with Nixon's visit to China. Reasons for yet another revival of interest are not difficult to find. Chinese anaesthetists, for example, were said by eyewitnesses to have carried patients through the ordeal of open-chest surgery by twirling a single needle in the patient's ear. And just in case anybody suspected the 'eyewitness' of exaggerating, the Chinese produced a film showing acupuncture anaesthesia in action. Many rational-minded clincans in the West tried to come up with an explanation and a possible mechanism for the effect. Heard most often was that acupuncture acted by releasing endogenous opioids (endorphins), those chemicals much hoped for by long-distance runners in the last few miles of a marathon to make their journey less painful.

But searching for a mechanism to explain acupuncture anaesthesia begs the more important question of whether the procedure works in the first place. Certainly in China there was far less enthusiasm for acupuncture after the fall of the Gang of Four, and in the 1980s two Chinese professors, Geng Xichen and Tao Naihuang, denounced surgical acupuncture in the Shanghai paper Wenhuibao as a myth and a political hoax. (Healthy scepticism has also somewhat dampened the enthusiasm that initially surrounded the discovery of endorphins. A typical comment: 'The opioid peptides appear to create euphoria not only in experimental animals but also indirectly in the investigators themselves.')

Does surgical acupuncture in particular, and acupuncture in general, work? The case for surgical acupuncture has all but been disproved. In Maoist China, surgical acupuncture was used in only 10% of carefully selected and prepared patients, with supplementary local and parenteral anaesthesia, and even then adequate anaesthesia was achieved in only one-third of these patients. What about the use of acupuncture in other areas where its use has been enthusiastically advocated by devotees—areas such as the control or cure of addictions (particularly smoking), reducing body weight, helping asthma sufferers, stroke rehabilitation, arthritis and back pain? (We shall ignore the more exaggerated claims for acupuncture made in Chinese publications—for example, that it cures deafness, schizophrenia, third-degree burns and malaria, despite the fact that these are often cited by acupuncturists in the West as support for the procedure.) Let's take a look at the evidence in the two cases of reducing body weight and helping smokers kick their habit.

Loosing Weight

In most societies before the twentieth century, and in many to this day, a wide girth has been viewed as a sign of health and prosperity. But more recently, corpulence has been seen as a growing threat to the health of many inhabitants of the richest nations. In the United States, for example, it is estimated that each year roughly 300,000 men and women are sent to early graves by the damaging effects of eating too much and moving too little. Fifty-nine per cent of the adult population of the USA meets the current definition of clinical obesity. A majority of women and many men are constantly dieting in an effort to control or reduce their weight and a variety of other treatments are in use—including acupuncture.

The use of acupuncture as a means of controlling appetite and reducing body weight arose from observations of individual cases—in essence, a series of anecdotes. But now a number of clinical trials have been undertaken to collect more reliable evidence. Some of these trials have been flawed (and even then gave contradictory results) but two relatively rigorous trials failed to support the efficacy of acupuncture. One of these, carried out by Dr. Allison and colleagues in New York, randomly allocated subjects to acupuncture treatment and to a 'sham' acupuncture procedure as a control condition. After 12 weeks the average weight loss was 1.28 kg in the acupuncture group and 0.63 kg in the placebo group; such a difference is quite compatible with the acupuncture and sham acupuncture having equal effects on weight loss. Similarly, there were no statistically significant differences between the two groups regarding fat loss or blood pressure reduction.

Several reports have appeared that have speculated about the mechanism by which acupuncture might lead to appetite suppression and weight loss. In the absence of convincing evidence that acupuncture really does achieve this aim, such speculation appears premature at best.

Stopping Smoking

> *It is now proved beyond doubt that smoking is one of the leading causes of statistics.*
> Fletcher Krebel

Smoking causes about 3 million deaths worldwide each year and about 15% of all cancers are attributable to smoking. Despite the protestations and 'weasel words' of the major tobacco companies, nicotine is a strongly addictive drug and most smokers, however well motivated, find that kicking the habit is extremely difficult. Rates of success in giving up smoking are very low when compared with the proportion of smokers who wish to stop and repeatedly try to do so. In recent years, the main aid for smokers wishing to stop has been nicotine replacement therapy by way of either nicotine patches or nicotine chewing gum. Most studies of these have found that even when abstinence for several months has been achieved, the risk of relapse remains high.

In their desperation to save themselves, many smokers have resorted to other methods to help them quit—including, of course, acupuncture. But it is only in the last few years or so that attempts have been made to assess the effectiveness of acupuncture treatment on smoking reduction. The results have been poor and there is little evidence that acupuncture is effective; the results seem to be even more disappointing than those obtained with both forms of nicotine replacement.

In the two areas of reducing weight and quitting smoking, there is little acceptable evidence that acupuncture is effective. Trying to hide or deny such failures undermines the efforts of acupuncturists to establish the procedure as effective for at least some conditions. And in other areas— for example, the prevention of nausea in patients with chronic pain—some clinical trial results appear to indicate that the procedure is effective. But the question still remains as to whether acupuncture is an elaborately disguised placebo that fulfils a need for mysticism and ancient ritual. Of course, even if acupuncture acts only as a placebo, it might be valuable, but a spade should be called a spade.

One danger when evaluating the clinical trial evidence for acupuncture is that research conducted in some countries (China, Japan, Hong Kong, and Taiwan) has been found to be uniformly favourable to acupuncture. One explanation is that the practitioners in these countries are more expert at applying the technique. But it is more likely that acupuncturists in China are already convinced about the effectiveness of the treatment and perhaps 'doctor' the results a little to make them favourable. Certainly one doesn't have to be a hard-nosed sceptic to raise doubts about taking the majority of these reports very seriously.

Perhaps the jury is still out on some aspects of acupuncture treatment, but the evangelical note stuck by acupuncturist, R.A. Dale at the founding convention of the American Association of Acupuncturists and Oriental Medicine, held in Los Angeles in 1981, seems entirely unmerited by the hard clinical trial evidence now available:

Acupuncture is a part of a larger struggle going on today between the old and the new, between dying and rebirthing, between the very decay and death of our species and our fullest liberation. Acupuncture is part of a New Age which facilitates integral health and the flowering of humanity.

No master of the understatement is Mr. Dale. And his suggestions for the strategy to be adopted by acupuncturists when dealing with the public border on the sinister:

- Undermine their faith in modern medicine and science
- Educate them in the need for alternative medicine
- Explain to then that what they need is not a medical specialist but an acupuncture generalist.

Homoeopathy

Homoeopathy is based on the principle of infinitesimally diluted 'remedies' that in higher dose produce the symptoms at which the treatment is directed. For example, arsenic, causes vomiting. The homeopath then takes a teaspoon of arsenic dilutes it with a hundred teaspoons of water and then repeats the process, say, six times. The diluted arsenic is then used as a treatment for vomiting.

It is the dilution principle that has caused so many scientists to be sceptical about the approach even before considering any empirical results collected from properly performed clinical trials. For homeopaths, the more diluted the solution, the more potent it becomes, if the process of imparting 'vital force' to the diluent by 'shaking' is properly carried out. The commonest homoeopathic dilution is roughly equivalent to one grain of salt dissolved in a volume of diluent which would fill 10,000 billion spheres, each large enough to enclose the whole solar system. The reason for prior scepticism is pretty clear. Unfortunately, homoeopaths have been reluctant to challenge their critics by organizing clinical trials, and sound, placebo-controlled studies are scarce. One of the oft-repeated mantras is that the homoeopathic treatments are 'highly individualized'. But some clinical trials have been attempted—let's examine three.

Warts and All

When I was about nine, I got a wart on the back of my hand. I seem to remember that having warts carried quite a stigma and so I felt pretty miserable. One day one of my brothers-in-law, who had been a sailor in World War II and so had a certain glamour in the eyes of an impressionable

9-year-old, claimed he could cure the unsightly blemish. The 'treatment' he prescribed was to spit on the wart before I spoke to anybody in the morning. This I did and, miracles of miracles, about 4 days later the wart had completely disappeared. Needless to say I have recommended this treatment ever since.

In 1996, a number of researchers carried out a randomized, double-blind clinical trial comparing individually selected homoeopathic preparations with placebo for the treatment of the common wart on the backs of children's hands. Thirty children received a homoeopathic remedy and thirty the placebo, the groups being formed by random allocation. The area occupied by warts was measured by computerized planimetry before and after 8 weeks of treatment. Reduction of the warty area by at least 50% was considered a successful outcome. In the group taking the homoeopathic treatment, nine were treated successfully; in the placebo group the corresponding figure was seven. These figures are quite compatible with the probability of a successful outcome being the same in each group.

Adenoid Enlargement

In 1997, a German research team undertook a randomized, double-blind clinical trial of the efficacy of homoeopathic treatment for children with adenoids enlarged enough to justify an operation. Patients were treated with either homoeopathic remedies (such as Nux vomica D200, Okoubaka D3, Tuberculinum D200, Barium jodatum D4 and Barium jodatum D6) or placebo. The duration of the study for each patient was 3 months. At the end of the study no operation was required in 70.7% of the placebo-treated children and in 78.1% of the children treated with homoeopathic preparations. The small difference is easily accountable for by the vagaries of random variation.

Migraine

Migraines are extremely painful headaches that can be debilitating. Only one side of the head is involved in many cases. Circulation to the scalp and brain can be altered, which affects the person's perception, muscle tone, and mental function, causing weakness, nausea, vomiting, sweating, chills and visual disturbances. A tendency toward migraines often runs in families, and allergic factors seem to be involved. Homeopaths recommend a number of remedies, belladonna, bryonia and cyclamen amongst them. All such remedies are said to reduce the pain and sickness associated with migraine, particularly if taken in the early stages of an attack. Sadly, these claims are not supported by the evidence from the few clinical trials of adequate methodological quality that have been undertaken. Such trials show no difference between homeopathy and placebo in the treatment of migraine.

The conclusion from the three studies described above and from others reported in the scientific literature leads to a clear conclusion: Homeopathy appears to be an elaborate method for the manufacture of placebos.

Summary

According to Skrabanek and McCormick, all systems of alternative medicine have two things in common. They have no detectable or coherent raison d'etre other than the enthusiasm of their advocates and, almost without exception, they claim to cure or alleviate a very large number of ill-defined and quite disparate ills. Some claim to have reached the Holy Grail of the cure-all. Many scientists and not a few clincians might, after studying the evidence from the clinical trials of alternative therapies that have been carried out, be inclined to agree with Skrabanek and McCormick. Such scepticism is summarized brilliantly in Catherine Bennett's article, 'No Alternative', published in The Guardian of Saturday, 15th August 1998. Some extracts follow:

> One thing we will not lack, in our nurse and doctor-less drift towards the grave, is complimentary therapy. When we die, we will do so with our feet massaged, our auras gleaming, our energies finely balanced, with a last drop of Dr. Bach's Rescue Remedy melting on out tongue. A recent report for the Department of Health showed that complimentary and alternative therapists outnumber Gps by around 40,000 to 36,300. Thanks to Simon Mills, the Director of Exeter University's Centre for Complimentary Health Studies, we know there are a minimum of 7,000 aromatherapists, 14,00 healers, 5,300 reflexologists, and 1,500 homeopaths to look after us, not to mention all the acupuncturists, chiropractors, osteopaths and rebirthers. Recruitment, Mills adds, is thriving.
>
> The most miraculous aspect of the rise and rise of alternative medicine, is that its popularity appears to be almost entirely unrelated to efficacy. While orthodox medicine has got bogged down in randomized trials, demands for evidence and cost-effectiveness...the shining reputation of alternative therapy is built on anecdotes...
>
> As Edzard Ernst, Professor of Complimentary Medicine at the University of Exeter, has warned recently, the debate 'has gradually become ill-informed, misleading and seriously unbalanced', with the media presenting individual yarns as proof of cure.
>
> As ever, leading the way in the cause of credulity is the Prince of Wales. Since 1982, when he chose the 150th anniversary of the BMA to reprimand doctors for their hostility 'towards anything unorthodox or unconventional', he has worked tirelessly to obtain alternative therapies on the NHS...

Some traditionalists have pointed out that it might be as well to check that complimentary medicine works, before it is integrated as the Prince suggests. But that was before the NHS staffing crisis. Can we now afford to be so picky about employing alternative practitioners? It would be fair, of course, for them to work in the same conditions as conventionally trained medics. They would have the GP's standard five minutes per patient, in which to treat the whole person. They would be on call at nights, and drafted into the casuality wards, especially on Saturdays, when the need for Dr. Bach's Rescue Remedy is at its greatest.

If the experiment succeeded, we would soon, no doubt, be seeing Aromatherapists Sans Frontieres taking their cures to the world's trouble spots. A homeopathic remedy based on scrapings of Canary Wharf—or should it be Norfolk, on the like-curing-like principle? could halve the Viagra bill. The alternative might well become proudly mainstream. Alternatively, it might just collapse and never be heard of again.

The Professor of Complementary Medicine mentioned in Caroline Bennett's article himself published an article in The Independent on 2nd June 1998. Professor Ernst's thoughtful, balanced and readable article summarizes the present situation regarding complementary medicine extremely well, and I can do no better than to conclude this chapter with his judgements on a variety of alternative therapies.

1. Examples of effective complementary treatments based on overviews of randomized clinical trials:

   - St John's Wort (hypericum), used in herbal medicine, alleviates the symptoms of mild-to-moderate depression and is associated with fewer short-term side effects than conventional drugs. But it should only be taken under clinical supervision because, according to Professor Ernst, 'A seriously depressed person who bought St. John's Wort might recover enough energy to commit suicide.'
   - Acupuncture reduces back pain, dental pain and nausea (e.g. morning sickness).
   - Ginkgo biloba, a herbal treatment, delays the clinical deterioration of patients with Alzheimer's disease.

2. Examples of forms of complimentary medicine that have been shown to be ineffective based on overviews of randomized clinical trials:

   - Acupuncture is no better than a sham as a help in smoking cessation or reduction of weight.
   - Iridology, diagnosis through the study of the iris, is not a valid diagnostic tool.
   - Chelation therapy, a vitamin infusion to cleanse the body of toxins, has been shown to be no better than placebo for circulatory problems in the leg.

3. Example of complementary treatments where, contrary to general beliefs, the evidence is inconclusive or insufficient based on overviews of randomized clinical trials:

- Chiropractic treatment has not convincingly been shown to be more effective than conventional treatments for acute or chronic low-back pain (nor for any other medical condition).
- Acupuncture has not convincingly been shown to be effective for arthritis or asthma.
- Hypnotherapy has not convincingly been shown to be more useful than standard therapies as an aid for smoking cessation.
- No evidence that herbal remedies for osteoarthritis and for the side effects of chemotherapy are effective.

4. Examples of complementary therapies which are associated with serious health risks:

- Chiropractic treatment can result in vascular damage (e.g. stroke) in an unknown number of cases.
- Acupuncture has been associated with serious internal injuries (e.g. collapsed lungs) and infections (e.g. hepatitis).
- Chelation therapy, colonic irrigation, certain diets (e.g. macrobiotic) have been linked to serious complications.
- Several herbal remedies are known to be toxic and can cause liver damage,others (e.g. Ayurvedic and Chinese remedies) have repeatedly been found to be contaminated with toxic substances.

5. Examples of therapies for which, so far, virtually no clinical trials have been carried out:

- Flower remedies/essences
- Shiatsu
- Crystal therapy
- Shark cartilage
- Rolfing (a deep form of massage guided by the contours of the body).

Debates about alternative medicine usually involve an extreme 'pro' or an extreme 'con' standpoint. Often emotion rather than reason is the driving force. There is nothing inherently wrong with promoting alternative therapies, but promotion should be based on reliable evidence rather than on the anecdotes and testimonials of users or therapists (When buying a used car, do you always believe the car dealer's claims that it was owned by one elderly lady and was used 'used only for short journeys to church on Sunday mornings'?) Contrary to what many supporters of complementary medicine like to maintain, it does not defy scientific testing. More statistically acceptable, properly designed, randomized, clinical trials are needed since as Professor Ernst has quite rightly pointed out, 'Proper testing of alternative remedies is ridiculously limited compared to the numbers of people who use them'.

Quackery has reached epidemic proportions, and the randomized clinical trial is the scientific cure.

What Does the Future Hold?

14

The stock market has forecast nine of the last five recessions.
Paul A. Samuelson

The future is hidden from us by infinite wisdom, or else I would like to know it; one would calculate one's behaviour at the present time so much better if one only knew what events were to become.
Mrs Gibson in *Wives and Daughters*, Elizabeth Gaskell

I don't believe in astrology. The only stars I can blame for my failures are those that walk the stage.
Noel Coward

Introduction

Do you remember that line from a Pink Floyd number, 'idling away the hours that make up a dull day'? I was reminded of it yesterday, an overcast, dull day in South London, nobody except the cats in the house, writing not going well, and even the music from the *Virgin Classic Rock* radio station failing to excite. I was about to resort to watching daytime television when I thought no, surely I'm not that desperate? Well not quite but nearly, since I did succumb to the somewhat double-edged pleasures of the Internet, and began some haphazard 'googling'. A number of medical sites quickly convinced me that I had a variety of unpleasant and possibly lethal conditions, and that my diet and exercise regimes were incompatible with a long and healthy life. I lingered a little on the gambling sites, passed rapidly over the sites peopled with rather unattractive couples (or more) in a variety of odd positions and activities, and finally in a flash of inspiration given that I was working on a chapter about predicting the future, I googled 'horoscope'. I looked at the first two out of about one thousand hits, in each case putting in my star sign, Gemini, and obtained two horoscopes for tomorrow. This is what I read:

B. Everitt, *Chance Rules*, DOI: 10.1007/978-0-387-77415-2_14,
© Springer Science+Business Media, LLC 2008

Hit 1

Today your creative powers will work well. Those born under your oppo-site sign, Sagittarius, may be particularly interested in your plans. Make sure to push your own ideas hard and let your voice ring out to convince others.

Hit 2

Let others speak their piece; keep your thoughts to yourself for a time. A good day for resting since ideas will not come easily but this day will be a stepping stone to something greater. There is a full moon and it is possible that issues related to the family will in some way be in the spotlight.

'Tomorrow' has now come and gone and I can simply reflect that only the cats were interested in listening to my ideas, as far as I know I did not bump into a Sagittarian, I'm still waiting for something greater to come about and my family arrived home at the same time and did the same things as on most other days.

As Niels Bohr rightly pointed out, 'prediction is very difficult, especially if it's about the future', and horoscopes and astrology in general are now regarded by most of us as bunkum, relatively harmless bunkum but bunkum nonetheless. Practising astrologers, of course, will continue to insist that our futures can be foretold in the stars, but personally until they can consistently tell me the winning lottery numbers week by week, they cannot expect to be taken seriously.

But if astrology is not the secret to knowing the future, what is? Are there scientific approaches to predictions that work? Or is prediction not only difficult but impossible? For answers to these questions read on.

Statistical Models

Statisticians often spend their time building models. But statistical models are not ship-in-a-bottle models or matchstick models of the Empire State Building. Rather, statistical models involve a series of mathematical equa-tions that purport to describe an activity or process and may help precise conclusions to be drawn about the activity or process and, in particular, may lead to predictions of what may happen in the future. Statistical models are, in some way, analogous to the 'ball-and-stick models' used by chemists to represent molecules and to mimic their theoretical proper-ties; such models can often be used to predict the behaviour of real objects.

As an example of a very simple statistical model, suppose a child has scored 20 points on some test of verbal ability and then, after studying a dictionary for some time, scores 24 points on a similar test. If we believe that studying the dictionary has led to the improvement then a possible model of what is happening is simply

$$20 = \{\text{student's initial score}\}$$

$$24 = \{\text{student's initial score}\} + \{\text{improvement}\}$$

So the improvement can be found by simply subtracting the first score from the second. But is this a sensible model for this situation? A little thought shows it is not, since it assumes that verbal ability can be measured exactly and this is unlikely to be the case—there will be some error of measurement each time it is measured. So a more realistic model of the two scores, one which allows for this measurement error would be

$$20 = \{\text{student's initial score}\} + \{\text{measurement error one}\}$$

$$24 = (\text{student's initial score}\} + \{\text{improvement}\}$$

$$+ \{\text{measurement error two}\}$$

Here, since we are allowing the amount of measurement error to differ when verbal ability is measured on each of the two occasions, we cannot get the *exact* improvement by simply subtracting the two verbal ability scores; under this model, subtracting the two scores gives us an estimate of the improvement, an *estimate* which will be 'close' to the true value if the measurement errors are small and not so close if they are large.

OK, this has all been serious stuff and, it has to be confessed, not that exciting. But if you're still awake let's cut to the chase and look at a situation where we want to use a statistical model for prediction. Enter the runners for the men's 1500m race in the Olympic Games.

Modelling Winning Times in the Men's 1500 m in the Olympic Games

The modern Olympics began in 1896 in Greece and have been held every 4 years since, apart from interruptions due to the two World Wars. On the track the blue ribbon event has always been the 1500m for men since competitors that want to win must have a unique combination of speed, strength and stamina combined with an acute tactical awareness. For the spectator the event lasts long enough to be interesting (unlike say the 100 m dash) but not too long so as to become boring (as do most 10,000 m races). The event has been witness to some of the most dramatic scenes in Olympic history; who can forget Herb Elliott winning by a street in 1960, breaking the world record and continuing his sequence of never being beaten in a 1500 m or mile race in his career? And remembering the joy and relief etched on the face of Seb Coe when winning and beating his arch rival Steve Ovett still brings a tear to the eyes of many of us.

The complete record of winners of the men's 1500 m from 1896 to 2004 is given in the table below. Can we use these winning times as the basis of a suitable statistical model that will enable us to predict the winning times for future Olympics?'

Herb Elliott winning the 1960 men's Olympic Games 1500 m (Credit: Topham Picturepoint, Edenbridge, Kent)

Seb Coe winning the 1980 men's Olympic Games 1500 m (Credit: Topham Picturepoint, Edenbridge, Kent)

Olympic Games, 1896–2004 winners of the men's 1500 m

Year	Venue	Winner	Country	Time(minutes and seconds)
1896	Athens	E. Flack	Australia	4.33.2
1900	Paris	C. Bennett	Great Britain	4.06.2
1904	St. Louis	J. Lightbody	USA	4.05.4
1908	London	M. Sheppard	USA	4.03.4
1912	Stockholm	A. Jackson	Great Britain	3.56.8
1920	Antwerp	A. Hill	Great Britain	4.01.8
1924	Paris	P. Nurmi	Finland	3.53.6
1928	Amsterdam	H. Larva	Finland	3.53.2
1932	Los Angeles	L. Beccali	Italy	3.51.2
1936	Berlin	J. Lovelock	New Zealand	3.47.8
1948	London	H. Eriksson	Sweden	3.49.8
1952	Helsinki	J. Barthel	Luxemborg	3.45.1
1956	Melbourne	R. Delaney	Ireland	3.41.2
1960	Rome	H. Elliott	Australia	3.35.6
1964	Tokyo	P. Snell	New Zealand	3.38.1
1968	Mexico City	K. Keino	Kenya	3.34.9
1972	Munich	P. Vasala	Finland	3.36.3
1976	Montreal	J. Walker	New Zealand	3.39.17
1980	Moscow	S. Coe	Great Britain	3.38.40
1984	Los Angeles	S. Coe	Great Britain	3.32.53

(Continued)

(*Continued*)

Year	Venue	Winner	Country	Time(minutes and seconds)
1988	Seoul	P. Rono	Kenya	3.35.95
1992	Barcelona	F. Cacho	Spain	3.40.12
1996	Atlanta	N. Morceli	Algeria	3.35.78
2000	Sydney	K. Ngenyi	Kenya	3.32.07
2004	Athens	H. El Guerrouj	Morocco	3.34.18

Let's begin by looking at a simple graph of the winning times against year for 1896–2004.

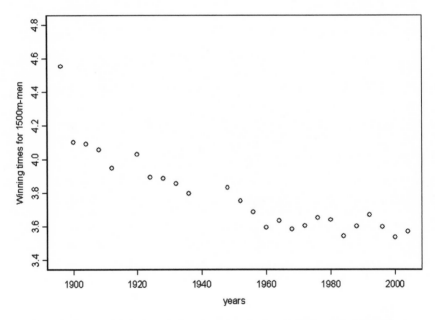

Plot of winning times in 1500m for men in the olympic Games against year

It is very clear from this graph that winning times are generally coming down over the years although the downward trend is not consistent; for example the winning time in 1984 was 3.32.53 but in 1992 it was far slower, 3.40.12. So, although the general direction of the times over the years is to go down, there are several little 'wiggles' in the graph. These wiggles reflect, essentially, the play of chance in the process; the appearance of a star runner leading to a winning time that bucks the trend by being too fast, for example, or a largely tactical race that leads to a slower time than might be expected.

There is one point on the graph which is way out-of-line with all the other winning times and this is the point representing the time for Mr. Flack in the 1896 games. Perhaps somebody forgot to tell the competitors they were allowed to run? But, whatever the reason for the very slow time, I am going to remove Mr Flack's time (this may look a little like 'fiddling' the data, but for statisticians, removal of an 'outlier' like Mr. Flack is meat-and-drink). Re-plotting the graph without Mr. Flack's time included gives the graph below:

This graph looks a much better bet for statistical modelling.

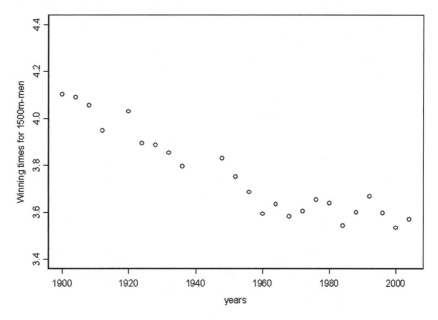

Times against year after removal of Mr. Flack

A very simple statistical model that we can use to describe the general tendency of the times to get faster over the years is to use the equation of a straight line to represent the relationship between winning times and years. We know, of course, that this will not describe the observations exactly because of the wiggles in the graph; a straight line model is only an approximation but it may still enable us to make some sensible predictions of times in future games. Such a model can be fitted very easily but the details of the fitting procedure can be ignored here and we can simply concentrate on the result, which is displayed in the next graph.

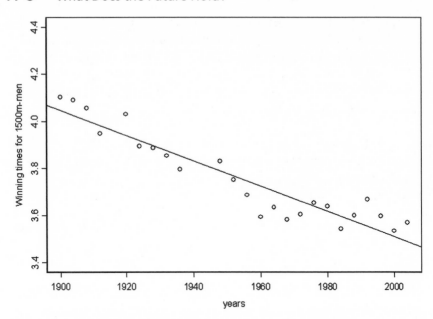

Times against year showing best straight line fit

The line shown on this graph is the 'best' (in some particular sense that we need not bother about) straight line for describing the relationship between the times and years from 1900 to 2004 and has the mathematical form:

$$\text{Winning time} = 14.21 - 0.0054 \times \text{Year}.$$

So, how can we use this equation to predict the times in future Olympics? We simply insert the year we are interested in, into the equation, do the simple arithmetic and find the corresponding time. For Bejing in 2008 and London in 2012 the calculations are:

Bejing-2008

Predicted winning time for 1500 m men $= 14.21 - 0.0054 \times 2008$
$$= 3.47 \text{ min, i.e. } 3.28.20 \text{ minutes and seconds}$$

London-2012

Predicted winning time for 1500 m men $= 14.21 - 0.0054 \times 2012$
$$= 3.45 \text{ min, i.e. } 3.27.00 \text{ minutes and seconds}$$

It's rather unrealistic to ask our model to predict an *exact* winning time; fairer is to ask for a *prediction interval* (statisticians would generally call these confidence intervals) which is very likely to contain the actual winning time. For 2008 and 2012 these intervals are

Prediction interval for 2008 = 3.24.89 − 3.31.36 minutes and seconds.

Prediction interval for 2012 = 3.23.43 − 3.30.25 minutes and seconds.

By the time this book is published readers will be able to check how the first of these two predictions has worked out.

The width of the prediction interval reflects the uncertainty in our prediction and will increase as we try to predict the winning time for games further into the future. For example, let's be very bold and use our model to predict the interval for the winning time for the year 2060. The result is;

Prediction interval for 2060 = 3.05.72−3.17.15 minutes and seconds.

Notice the width of the interval is considerably wider than when predicting the 2008 and 2012 times.

Now an important question that needs to be asked about any statistical model is whether it is realistic for the situation it purports to describe. So let's consider that question for our model for the winning times in the men's 1500 m. And it doesn't take a rocket scientist to see that our model is not altogether realistic for describing the progress of 1500 m times. This is easily illustrated if we try to use it to predict winning times far into the future; for example, for the year 2652 the predicted winning time is 1.38 SECONDS! Not much fun anymore for the spectators—blink and you've missed it! And what would the 100 m now be won in?

Now we can either believe that by this time there will be a species of supermen who can run at this speed or we can suspect that something isn't quite right about our model. I think we should suspect the model and an obvious problem with it is that it allows the winning times to continue to decrease without limit and this is clearly wrong. Even a basic knowledge of human physiology tells us that there must be some lower limit to the time in which a man can run 1500 m.

Our straight line model is just too simplistic for the situation we are modelling to give sensible predictions way into the future. The straight line model is probably not too bad for predicting a little way ahead, say to 2020, but I wouldn't bet my life on its predictions much further into the future than that. This illustrates one of the ever present dangers when using any statistical model for prediction, namely that trying to extrapolate too far outside the range of the observations on which the model is based can be dangerous.

Nobody is going to get rich predicting winning times in the Olympics. So let's continue our consideration of the use of statistical models where they might be more lucrative and that is helping us (perhaps) to make a killing on the stock market.

Can Statisticians or Mathematicians Predict the Market?

For those readers who took in Chapter 7 on their way here, it will be clear that on occasions I am not averse to a little gamble on the horses or at the casino. I also invest in the stock market both directly and indirectly through pension contributions. Note the curious niceties of language here; for horse racing and the casino I use *gamble* defined in my dictionary as:

> *to bet or risk something on an uncertain outcome,*

but for the stock market I use *invest* defined in the same dictionary as:

> *to commit money to a particular use in order to earn a financial reward.*

Is the distinction merited? Well I guess it depends if your chosen horse wins the 3.30 at Ascot or the roulette wheel stops on red to match your bet and your shares take a tumble, or whether your horse is still running long after the winner has finished, the wheel shows black and your shares go through the roof. One of the reasons perhaps that 'invest' is used in respect of the stock market is that stockbrokers have an air of respectability and disinterest that is not generally afforded to bookmakers or casino operators. Perhaps if stockbrokers were more commonly known as *share salesmen* then their views, such as 'the recent fall in share prices is a useful correction' would not be treated so differentially—useful for whom we might then ask?

But the distinction between gamble and invest might be merited if share movements could be predicted in some way since we know from what was written in Chapter 7 that prediction in roulette is impossible when there is a properly balanced wheel and is notoriously difficult (most punters would tend to go with impossible here as well) in horse racing. If some type of statistical/mathematical model can predict, even approximately, share prices over the medium term then using the word 'invest' rather than the word 'gamble' *would* be justified.

Well, many people have claimed that share prices *can* be predicted and there is no shortage of statistical models that have been suggested for the task. Here are just a few:

- Simple moving average
- Triangular moving average
- Andrew's pitchfork

- Fibonacci fan
- Ichimoku charts
- Adaptive stochastic oscillator.

Most of these are based on using past prices as the basis of what will happen in the future in a manner similar to that used in the previous section for the winning times in Olympic 1500 m races. Of course, many of the suggested models also have the addition of lots of 'bells and whistles' with the aim of making predictions more accurate. And some models for stock price prediction, of course, are far more sophisticated for example, the so-called 'super-intelligent' models involving chaos theory, neural networks and self-learning approaches.

(We can also note en passant that there are also astrological funds where the astrologer uses the movements of the heavens to predict the market, by looking up the 'birthdate' when companies were formed and then drawing up an astrological chart—enough said I think.)

One of the attractions to bright statisticians and mathematicians who attempt to model movements in share prices (apart from the obvious one of trying to make money) is that there is a mass of data about stock markets. With modern computing methods the interested statistician or mathematician can build their pet models from a gigantic database of every sort of economic statistic available; they are seduced into thinking that with so much data around producing a worthwhile model that can be used to make their fortunes is only a matter of time. But being smart might not make them rich, as two well-known failures of stock market modelling suggest:

- The Long Term Capital Management hedge fund which assembled awesome talent, including two Nobel Prize winners, to devise its super-duper computer model to predict market movements and bet millions (I'm sorry, invest millions) trying to anticipate future movements in the world's financial markets using options and futures investments, but were caught out when the fund's investments in Russian markets were hit by the rapid devaluation of the rouble. In 1999 the fund lost $2.3 billion and went bust.
- Goldman Sachs has become renowned as being one of the smartest players on Wall Street. But in August 2007 the bank had to admit that a flagship global equity fund had lost over 30% of its value in a week because of its problems with the models used. Goldman Sachs operated a computer model, devised by experts, for predicting future values and opportunities. The model allowed for everything—except what actually happened—over a billion dollars were lost in a flash. And as David Evans comments on his *Unqualified Remarks* blog, 'the problem is that when computers lose you a billion, they don't do the decent thing and jump out of the window, and firing them isn't as emotionally satisfying.'

Those statisticians and mathematicians who are investing their talents and time in trying to find some ultimate model for predicting share prices might do well to read John Allen Paulos's excellent and entertaining book, *A Mathematician Plays the Market*. Paulos decided to try to find out if a maths expert could outwit the stock market and make millions, by taking a gamble on the investment game himself. His book gives a witty and absorbing account of how he discovered that no amount of clever maths can guarantee you a fast buck on the stock market, that technical analysis simply does not work and that, in the market, randomness and chance rule along with a considerable dash of psychology. It seems likely that the movement of the stock market is almost entirely random; there may be a very long trend in the upward direction but to make money in the short to medium term you might (would) do better with a blindfold, pin and a list of shares than squinting at squiggles on charts, examining the ups and downs of a time series, and trying to extract a trend from moving averages of whatever sort. And even neural networks may not be up to the task of making accurate predictions when faced with stock prices.

So, perhaps I should now change the beginning of an earlier sentence in this section to 'I also gamble on the stock market...'.

So, to summarize:

- Statistical modelling can be helpful in some situations for making predictions but enormous care is needed when building the model to ensure that it is sensible for the situation or process being modelled.
- For any statistical model, trying to make predictions way outside the range of the data on which the model is constructed is often a complete waste of time.
- In situations as complex as the stock market the vast variety of factors affecting stock prices may be impossible to build into any model and make the movements in share prices largely random. Who, for example, could have predicted the terrible events of 9/11 which affected the market so dramatically? In such situations, prediction becomes almost as close to impossible as predicting the spin of a roulette wheel. Investing then becomes a euphemism that attempts to put a positive spin on an activity which is essentially gambling.
- A direct consequence of the previous bullet point is that there is no system that is capable of reliable and accurate prediction when it comes to stock market movements. And if anybody tries to tell you otherwise, walk away.

I can't end this chapter on foretelling the future without some mention of my favourite predictions that didn't quite work out:

It will be all gone by June.

Variety magazine passing judgement on rock and roll in 1955

It will be years, not in my time, before a woman becomes Prime Minister.

Margaret Thatcher, 1969

Heavier-than-air flying machines are impossible.

Lord Kelvin, 1895

There is no reason why anyone would want a computer in their home.

Ken Olsen, founder of DEC in 1977

Stocks have reached what looks like a permanently high plateau.

Irving Fisher, economics professor, Yale University, 1929.

And finally two predictions of my own about which I am quietly confident:

- Statisticians will soon become cult heroes, they will appear regularly on reality TV, will be hounded by the paparazzi on their way to meetings and have their private lives reported in *Heat* and *OK* magazine.
- England will thrash Australia in the next Ashes cricket series in 2009.

(Both predictions may be classic examples of the triumph of hope over experience—but we shall see.)

Chance, Chaos and Chromosomes

Everything existing in the Universe is the fruit of chance...
Democritus

Not only is there no God, but try getting a plumber at weekends.
Woody Allen

'Forty-two,' said Deep Thought, with infinite majesty and calm.
Douglas Adams, *The Hitch Hiker's Guide to the Galaxy*

The Overthrow of Determinism

Now that we have spent so many pages discussing the operation of chance in a variety of situations, the time has come to confront the difficult question, conveniently ignored up to now. Just what is chance? Cicero was ahead of his time when he characterized chance events as events which occurred or will occur in ways that are uncertain—events that may happen, may not happen, or may happen in some other way. Deborah Bennett in her excellent little book, *Randomness*, points out that the more widespread belief at the time was that what we call chance is merely ignorance of the determining cause of the event. According to Leucippus (*circa* 450 B.C.), the first atomist; 'Nothing happens at random; everything happens out of reason and by necessity'.

Leucippus thus associated the operation of chance with some *hidden* cause and the Stoics took a similar stance some 200 years later. The following is attributed to Chrysippus (*circa* 280–207 B.C.); 'For there is no such thing as lack of cause, or spontaneity. In the so-called accidental impulses which some have invented, there are causes hidden from our sight which determine the impulse in a different direction.

This deterministic view of the world, involving natural laws of cause and effect, continued to prevail during the Middle Ages in Europe and was strongly encouraged by the early Christian church's belief that everything

B. Everitt, *Chance Rules*, DOI: 10.1007/978-0-387-77415-2_15,
© Springer Science+Business Media, LLC 2008

happened at the will of the Creator (though strictly speaking, the latter perspective should be labelled fatalism). The English philosopher, Thomas Hobbes (1588–1679), for example, left no room for chance in his universe; he held that all events were predetermined by God or by extrinsic causes determined by God. Hobbes argued that it is ignorance of those causes that induces some people to overlook the necessity of events and attributing them instead to chance.

Acceptance of Christian determinism was reinforced by one of the greatest scientific revolutions in history, Newtonian physics. According to Bennett:

> based on the work of Newton and many others, a belief developed among scientists that everything about the natural world was knowable through mathematics. And if everything conformed to the design of mathematics, then a Grand Designer must exist. Pure chance or randomness had no place in this philosophy.

The doctrine of classical determinism at its most grandiose is reflected in the French mathematician Pierre-Simon Laplace's words in his *Analytical Theory of Probability* (1820):

> An intellect which at a given instance knew all the forces acting in nature, and the position of all things of which the world consists—supposing the said intelligence were vast enough to subject these data to analysis—would embrace in the same formula the motions of the greatest bodies in the universe and those of the slightest atoms; nothing would be uncertain for it, and the future, like the past, would be present to its eyes.

This view of nature reflects the laws of mechanics as developed by Galileo and Newton: The future motion of a particle can be predicted if we know its position and velocity at a given instant. More generally, the future behaviour of a system can be predicted from knowledge of all the requisite initial conditions. The world from this standpoint is a global automaton, without any freedom, determined from the beginning. Events may appear to us to be random but this could be attributed to human ignorance about the details of the processes concerned. Apparently chance events (such the toss of a coin, the roll of a die, or the spin of a roulette wheel) are accounted for by assuming that they would no longer be random if we could observe the world at the molecular level. Statistical ideas *were* accepted and used in nineteenth-century physics, (for example, in kinetics and statistical thermodynamics), but only when information was incomplete or because of the complexity of the system involved. The underlying laws were still considered to be deterministic, however. The cosmic clockwork machinery ensured that even the seemingly most haphazard events obeyed its laws, albeit in the most fearsomely convoluted manner.

Today, in the early part of the twenty-first century, any physics student now knows that the picture Laplace painted is wrong. Several events early in the twentieth century made the introduction of randomness and chance into physics a necessity. One of the first involved a direct experimental situation which focused attention on randomness. A series of experimental studies on alpha particles emitted from radioactive elements led Rutherford to state in 1912, 'The agreement between theory and experiment is excellent and indicates that the alpha particles are emitted at random and the variations accord with the laws of probability'.

What Rutherford had found was that the moment of emission of an alpha particle from a piece of radioactive material cannot be predicted—it is in essence unknowable—and thus there is an element of genuine unpredictability in nature. But Rutherford's work also showed that long-term properties of the process, such as the number of particles emitted in a particular time, did follow the pattern predicted by probability theory.

The development, in the 1920s, of quantum mechanics drove another nail in the coffin of classical determinism. A central tenent of this approach is that it is impossible to predict exactly the outcome of an individual measurement of an atomic (or molecular) system. The uncertainty here has nothing to do with the possibility of measurement error or experimental clumsiness; it is a fundamental aspect of the basic physical laws themselves. In the microworld of electrons, photons, atoms and other particles, paths travelled can only be described probabilistically. When a bullet arrives at a target, its point of arrival represents the end point of a continuous path that started at the barrel of the gun. Not so for electrons. We can discern a point of departure and a point of arrival, but we cannot always infer that there was a definite route connecting them—a paradox encapsulated in the famous uncertainty principle of the German physicist Werner Heisenberg. The probabilistic nature of quantum theory can be illustrated by a simple example taken from Murray Gell-Mann's book, *The Quark and the Jaguar*:

> A radioactive atomic nucleus has what is called a 'half-life,' the time during which it has a 50% chance of disintegrating. For example, the half-life of Pu^{239}, the usual isotope of plutonium, is around 25,000 years. The chance that a Pu^{239} nucleus in existence today will still exist after 25,000 years is 50%; after 50,000 years, the chance is only 25%; after 75,000 years, 12.5%; and so on. The quantum-mechanical character of nature means that for a given Pu^{239} nucleus, that kind of information is all we can know about when it will decay; there is no way to predict the exact moment of disintegration. *cdots* And the direction of decay is completely unpredictable. Suppose the Pu^{239} nucleus is at rest and will decay into two electrically charged fragments, one much larger than the other, travelling in opposite

directions. All directions are equally likely for the motion of one of the fragments, say the smaller one. There is no way to tell which way the fragment will go.

Gell-Mann then reflects, 'If so much is unknowable in advance about one atomic nucleus, imagine how much is fundamentally unpredictable about the entire universe'.

"WHAT'S COME OVER HEISENBERG? HE SEEMS TO BE CERTAIN ABOUT <u>EVERYTHING</u> THESE DAYS."

Given its probabilistic interpretation of the wave function and its associated unaskable questions, it is not surprising that Niels Bohr once remarked that anybody who is not shocked by quantum theory has not understood it. But thanks to its ability to explain a wide range of otherwise incomprehensible phenomena, quantum theory is now often cited as the most successful scientific theory ever produced. Its success has *probably* forever doomed the deterministic picture of the universe.

Chaos

Quantum theory incorporates what we might call *true* randomness in nature. But even when it is scrutinized independently of this true randomness, classical determinism contains *within itself* a bankruptcy that renders Laplace's picture useless. There are some phenomena whose evolution exhibits extreme sensitivity to the starting state. The slightest change in the initial state results in an enormous difference in the resulting future states. Minute changes to the initial state quickly destroy the practical utility of the deterministic picture even as an approximation and the behaviour of the phenomena is then, in all practical respects, random and unpredictable. An example is a large torrent of turbulent fluid. Such a fluid might start with a fairly uniform flow, different parts of it being very similar in speed and direction of motion. Yet, after it falls over a waterfall, say, these small initial differences between the water motions at neighbouring places become enormously amplified.

James Clerk Maxwell (1831–1879) was perhaps the first to appreciate that there is a world of difference between determinism in principle and determinism in practice:

> When the state of things is such that an infinitely small variation of the present state will alter only by an infinitely small quantity the state at some future time, the condition of the system, whether at rest or in motion, is said to be stable; but when an infinitely small variation in the present state may bring about a finite difference in the state of the system in a finite time, the condition of the system is said to be unstable.
>
> It is manifest that the existence of unstable conditions renders impossible the prediction of future events, if our knowledge of the present state is only approximate, and not accurate....

Maxwell was essentially drawing the attention of his colleagues to systems in which a minute uncertainty in their current state prevents us from accurately predicting their future state. Only if the initial state were known with perfect accuracy would the mathematical equations of classical, deterministic physics be of use. But perfect accuracy is not attainable either in practice (of which Maxwell was well aware) but also in principle, because, as we now know (but Maxwell did not), the quantum aspects of reality forbid the acquisition of such error-free knowledge of the initial conditions. Nor are these quantum restrictions far removed from experience as John Barrow points out in his fascinating book, *Theories of Everything*:

> If we were to strike a snooker ball as accurately as the quantum uncertainty of nature permits, then it would take merely a dozen collisions with the sides of the table and the other balls for this uncertainty to have amplified to encompass the

extent of the whole snooker table. Laws of motion would henceforth tell us nothing about the individual trajectory of the ball.

In other words at the end of the dozen collisions, application of the equations of motion would allow us to say that the ball is on the table and nothing else.

(Here I cannot resist bringing in our old friend, Galen,who even earlier than the great Maxwell, recognized situations where the effect of a cause is disproportionate and evident: 'In those who are healthy...the body does not alter even from extreme causes; but in old men even the smallest causes produce the greatest change'. Unfortunately, I know exactly what Galen was getting at.)

Nowadays, deterministic processes that exhibit apparently random behaviour are collected together as so-called chaotic phenomena, and chaos theory has, in the last decade or so, become as fascinating for the general public as to professional mathematicians. The topic has spawned a multitude of popular accounts, of which the best by some distance are those of Ian Stewart (*Does God Play Dice?*) and James Gleick (*Chaos*). The interplay of determinism and chaos has become so fashionable a topic that even modern playwrights expect their West End and Broadway audiences to take an interest. And so, nearly 200 years after Laplace's remarks, two characters in Tom Stoppard's play *Arcadia* ponder thus on determinism and its downfall:

> CHLOË: The future is all programmed like a computer —that's a
>   proper theory, isn't it?
> VALENTINE: The deterministic universe, yes.
> CHLOË: Right. Because everything including us is just a lot of atoms
>   bouncing off each other like billiard balls.
> VALENTINE: Yes. There was someone, forgot his name, nineteenth
>   century, who pointed out that from Newton's laws you could predict
>   everything to come—I mean, you'd need a computer as big as the universe
>   but the formula would exist.
> CHLOË: But it doesn't work, does it?
> VALENTINE: No. It turns out the maths is different.

This mathematics that is 'different' is the mathematics of chaos theory, which says that a situation can be both deterministic *and* unpredictable. Chaos provides a bridge between the laws of physics and the laws of chance.

Evolution, Genes and Chromosomes

Up to the middle of the nineteenth century, the conventional biological view was that all species were immutable and that each had been individually and separately created by God. Pre-Darwinian scientists had made detailed statistical studies that reveal some belief in mathematical laws governing the development and variation of living things. They recognized that there is, at the root of reproduction, a statistical element whose uniformity in the long run requires an explanation. However, the explanation sought was not a scientific one but involved some version of the Grand Design. This view was shattered in 1859 with the publication of Charles Darwin's *On the Origin of the Species by Means of Natural Selection or the Preservation of Favoured Races in the Struggle for Life*, the first account of Darwin's theory of evolution, now almost universally acknowledged as the major intellectual event of the nineteenth century.

It was during his 5 years aboard the *HMS Beagle* on a surveying expedition to Patagonia that Darwin began to suspect that species are not fixed forever. Using clues from an essay on population by Malthus, and drawing on evidence from his observations during his trip, he formulated his theory of evolution. Briefly put, his argument was that all individuals of the same species are not identical—organisms vary. Those organisms survive that have the variations that best fit them to thrive and reproduce in the environment they inhabit, and these successful variations are passed on to progeny. An example provided by David Attenborough in his *Life on Earth* serves to illustrate Darwin's ideas more colourfully:

> In one clutch of eggs from, for example, a giant tortoise, there will be some hatchlings which, because of their genetic constitution , will develop longer necks that others. In time of drought they will be able to reach leaves and so survive. Their brothers and sisters, with shorter necks, will starve and die. So those best fitted to their surroundings will be selected and be able to transmit these characteristics to their offspring. After a great number of generations, tortoises on the arid islands will have longer necks than those on the watered islands. And so one species will have given rise to another.

The random element of evolution is in the numerous, successive, slightly favourable variations that produce offspring that differ slightly from the parents—a biological example of the random variation encountered in earlier chapters. Darwin, lacking an acceptable theory of heredity, had little conception of how these variations came about. But, like the physicists of the day, he tended to believe that the apparent operation of chance masked the exact laws that existed but that, as yet, were unknown: 'I have hitherto sometimes spoken as if the variations...had been due to chance. This, of course, is a wholly incorrect expression, but it seems to acknowledge plainly our ignorance of the cause of each particular variation'.

Others were equally dismissive of a chance mechanism being a component of evolution. Here is the famous astronomer, Sir John Herschel, for example:[*]

> We can no more accept the principle of arbitrary and casual variation and natural selection as a sufficient account, *per se*, of the past and present organic world, than we can receive the Laputian method of composing books...as a sufficient one of Shakespeare and the *Principia*.

But when a usable theory of heredity was developed during the next half-century, randomness played a major role, both in errors in copying the genetic message—*mutations*—and in the genetic inheritance of offspring. In spite of Darwin's reservations, chance became increasingly important in evolutionary theory. As Hopkins succinctly put it in his *Chance and Error—The Theory of Evolution*, 'The law that makes and loses fortunes at Monte Carlo is the same as that of evolution'.

Darwin's ideas presented the living world as a world of chance, determined by material forces, in place of a world determined by a divine plan. The substitution of chance for purpose was what caused such a furore when the ideas were first published. Richard Dawkins eloquently expresses the paradox and summarizes:

> Natural selection is the blind watchmaker, blind because it does not see ahead, does not plan consequences, has no purpose in view. Yet the living results of natural selection overwhelmingly impress us with the appearance of design as if by a master watch maker, impress us with the illusion of design and planning.

Darwin himself was terrified by his own ideas. Although by the late 1830s he was convinced that humankind and the other animal species had not been separately created by God, but rather had evolved from a common ancestor, he refrained for 20 years from publishing his epic work on the origin of the species. Only when he discovered that a younger biologist, Alfred Russel Wallace (1823–1913), had reached similar conclusions independently did he steel himself and go public. His fear was that the blow to Christianity and to human dignity dealt by his evolutionary theory would encourage atheistic agitators and socialist revolutionaries. Darwin was wedded to respectability and social order, although by now he did not believe in an omnipotent God.

Darwin's fears about the possible conflict between his theory of evolution and Christianity have been proved to be well founded. In the early part of the twenty-first century it appears that the majority of Christians are

---

[*] [Herschel is suggesting here that The Literary Engine observed by Gulliver on his visit to Lilliputia was unlikely to ever produce a Shakespeare play. Essentially the Engine put words into sentences at random.]

extremely anti-evolution. Some evidence for this claim is that the United States surveys show that 52% of the population do not believe Darwin's theory of evolution. The popular alternative belief to evolution, creationism, is that the universe and everything in it was created some 6000 years ago by 'God'. Creationism conflicts with almost all other scientific theories, for example, plate tectonics, the fossil record, carbon dating using radioactivity, and most of biology. The truth of creationism would imply the necessity for a pretty radical overhaul of all areas of science. And compared to the elegance and beauty of Darwin's theory of evolution (forgetting for a moment that it's also very likely to be true), the creationism theory is damned by being dull and boring (forgetting for a moment that it's almost certainly not true). And how do creationists explain what their god was about when he created something as useless as the wasp and as ugly as the stone fish? If you want a more detailed scientific argument of why evolution is right and creationism and its attempted more respectable, but actually equally absurd cousin, intelligent design are wrong, you can do no better than read the brilliant book by Sean Carroll, *The Making of the Fittest.*

In a country as great as America it's sad to find seven candidates for the Republican Presidential nomination publicly denouncing evolution in favour of creationism knowing that if they don't, they can say goodbye to any chance of being elected. It appears that it will be some time before a Presidential candidate for either party will be able to declare a belief in evolution along with a disbelief in god and still be nominated.

The theory of heredity that Darwin lacked to explain the random variation in species was developed only after his death. The concept that human characteristics are transmitted from parents to offspring in discrete units called *genes* is now so well established that it may appear self-evident to many people. In fact, up to the early part of the last century, the gene concept was virtually unknown, and most experts believed in a theory of inheritance called *pangenesis.* This theory supposes that each sperm or ovum contains minute quantities of every organ of the parent and that the organs of the offspring develop from these raw materials. This theory can be traced back as far as Hippocrates (about 400 B.C.), who said that 'The seeds come from all parts of the body, healthy seed from healthy parts, diseased seed from diseased parts'.

It was Gregor Mendel's investigation of garden peas that provided the first convincing evidence of the existence of genes (although there is some concern that Mendel was not totally honest in the reporting of his results), and the gene concept became popular in the early 1900s. At this time, the gene was essentially a theoretical construct useful in explaining heredity, much like the atom had been in chemistry several decades previously. But, by the mid-1930s, it had been established that genes were physical entities and eventually in the 1950s, the structure of the gene was discovered. I'll let Francis Crick briefly continue the story. The following is from his auto-biography, *What Mad Pursuit*:

> At the time I started in biology—the late 1940s—there was already some rather indirect evidence suggesting that a single gene was perhaps no bigger than a very large molecule—that is, a macro-molecule. Curiously enough, a simple, suggestive argument based on common knowledge also points in that direction.
>
> Genetics tells us that, roughly speaking, we get half of all our genes from our mother, in the egg, and the other half from our father, in the sperm. Now, the head of a human sperm, which contains these genes, is quite small. A single sperm is far too tiny to be seen clearly by the naked eye, though it can be observed fairly easily using a high-powered microscope. Yet in this small space must be housed an almost complete set of instructions for building an entire human being (the egg providing a duplicate set). Working through the figures, the conclusion is inescapable that a gene must be, by everyday standards, very, very small, about the size of a large chemical molecule. This alone does not tell us what a gene does, but it does hint that it might be sensible to look first at the chemistry of macromolecules.

By the early 1950s it was known that the chemical material of the gene was DNA (deoxyribonucleic acid), and in 1953 Watson and Crick made one of the most famous breakthroughs in modern science with their discovery of the structure of DNA. The DNA molecule, they suggested, is in the form of a double helix, with two distinct chains wound round one another about a common axis. Crick and Watson also suggested that to replicate DNA, the cell unwinds two chains and uses each as a template to guide the formation of a new companion chain—thus producing two double helices, each with one new and one old chain.

The genes are physically located in the chromosomes in the cell nucleus. A chromosome is a thread-like structure consisting of proteins and small amounts of DNA. The genetic blueprint of an individual is contained in 23 pairs of chromosomes. The fertilised egg and the adult that grows from it contains an essentially random mixture of characters from each parent. The action of the chromosomes when sperm encounters egg is governed by chance. And the chance mutations that occur in this process are the driving force behind evolution, Darwin's apparent distaste for such a mechanism notwithstanding.

Thus life is, quite literally, a gamble—a situation that is masterly sum-marized by Richard Fortey in one of the best popular science books of recent years, *Life: An Unauthorised Biography*:

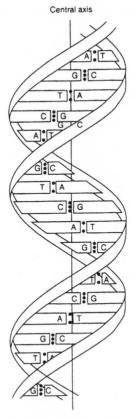

The structure of DNA. The two bands represent the sugar phosphate backbones of the two strands, which run in opposite directions. The vertical line represents the central axis around which the strands wind. The positions of the four nucleotide bases (C, A, T and G) are shown together with the hydrogen bonds (●) which link them together

Chance promotes or damns according to the whim of history. The analogy of permutations in a game of chance has long been applied to genetic mutations, spontaneous changes in the genetic code. Like the vast ranks of gamblers in Las Vegas who come away with nothing, most mutations also lead to nothing. They do not result in an increase in fitness. Some are lethal—as might be the production of a wingless butterfly. Others might put the mutant at a disadvantage—a new colour pattern might not be favoured by a potential mate, for example. But those rare mutations that hit upon an advantageous combination produce the big payoff—the jackpot. They are rewarded not in gross currency of quarters or dollars but in the irresistible coin of many successful offspring. Unlike the lucky gambles in Excalibur or Caesar's Palace the luck of the successful gene is passed on to make luck for future generations.

Life, the Universe and Everything

> On Friday afternoon, July the twentieth, 1714, the finest bridge in all Peru broke and precipitated five travellers into the gulf below.

Thus begins Thornton Wilder's 1927 novel, *The Bridge of San Luis Rey*. A witness to the accident is the 'little red-haired Franciscan from Northern Italy', Brother Juniper, who happens to be in Peru converting the Indians. Brother Juniper, eager for theology to take its place among the exact sciences and accepting that the collapse of the bridge of San Luis Rey was a sheer Act of God, resolves to investigate the question, 'Why did it happen to *those* five?' He believes that if there were any plan in the universe at all, if there were any pattern in a human life, surely it could be discovered mysteriously latent in those lives so suddenly cut off. Either we live by accident and die by accident, or we live by plan and die by plan. To Brother Juniper there was no doubt; he knows the answer. He merely wants to prove it—historically, mathematically—to his converts.

For the next 6 years he keeps himself busy knocking on all the doors in Lima, asking thousands of questions, filling scores of notebooks, in an effort to justify the ways of God to man. The result of all Brother Juniper's diligence is an enormous book, dealing with one after another of the victims of the accident, cataloguing thousands of little facts and anecdotes and testimonies, and concluding with a dignified passage explaining why God had settled upon that person and upon that day for His demonstration of wisdom. Brother Juniper was clearly involved in a heroic effort to prove that old Christian apologist's aphorism 'God works in a mysterious way his wonders to perform'. Sadly, Brother Juniper comes to learn that investigating the mind of God can be a chancy business—his book eventually comes to the attention of some judges and is pronounced heretical. Both the book and the author are burned in the town Square.

When Loonquawl and Phouchg, two characters in Douglas Adams's cult book of the 1970s, *The Hitch Hiker's Guide to the Galaxy*, interrogated the computer Deep Thought as to the answer to the great question of life, the Universe and everything, they received the reply given in one of the quotations that opened this chapter. Deep Thought was apologetic, 'You're really not going to like it' he counselled. But perhaps 'forty-two' is about as good an answer that those bold enough to seek to uncover the ultimate mystery of existence should expect. Personally I suspect Deep Thought got it slightly wrong. His answer should have been 'probably forty-two' because, as John Barrow observes at the end of *Theories of Everything*, 'There is no formula that can deliver all truth, all harmony, all simplicity. No Theory of Everything can ever provide total insight. For, to see through everything, would leave us seeing nothing at all.'

It seems that Deep Thought should have shown a little more humility!

But despite Barrow's reservations, philosophers, physicists and theologians continue to speculate on the big questions and, in particular, on the biggest: just how the Universe came into being. According to Greek mythology, three brothers rolled dice for the Universe, Zeus winning the Heavens, Poseidon the seas, and Hades (the loser) going to hell as master of the underworld. But the standard twenty-first century explanation is the Big Bang theory—all matter and energy in the universe originated in a superdense agglomeration that exploded at a finite moment in the past. The theory successfully explains the expansion of the universe, the observed microwave background radiation, and the observed abundance of helium in the universe. Tell this to one's non-scientist friends, however, and many of them come back with further questions. 'What was it like before the Big Bang?' or 'where did the material for the Big Bang come from?' They are, justifiably perhaps, less than satisfied with the standard answer that this is no longer the realm of science and that physics has nothing to say about what preceded the Big Bang. Most physicists suggest that the phrase *before the Big Bang* has no meaning, because the flow of time did not exist. Even more damning perhaps is that they don't find the question of any interest. Most but not all. The following is taken from Alan Guth's book *The Inflationary Universe*:

> In the late 1960s, a young assistant professor at Columbia University named Edward P. Tyron attended a seminar given by Dennis Sciama, a noted British cosmologist. During a pause in the lecture, Tyron threw out the suggestion that 'maybe the universe is a vacuum fluctuation'. Tyron intended the suggestion seriously, and was disappointed when his senior colleagues took it as a clever joke and broke into laughter. It was, after all, presumably the first *scientific* idea about where the universe came from.
>
> By a *vacuum fluctuation*, Tyron was referring to the very complicated picture of a vacuum, or empty space, that emerges from relativistic quantum theory. The hallmark of quantum theory, developed to describe the behaviour of atoms, is the probabilistic nature of its predictions. It is impossible, even in principle, to predict the behaviour of any one atom, although it is possible to predict the average properties of a large collection of atoms. The vacuum, like any physical system, is subject to these quantum uncertainies. Roughly speaking, *anything* can happen in a vacuum, although the probability for a digital watch to materialize is absurdly small. Tyron was advancing the outlandish proposal that the entire universe materialized in this fashion!

Ten years later, Tyron returned to his idea publishing in *Nature* a paper entitled,' Is the Universe a Vacuum Fluctuation?' His crucial point was that the vast cosmos we see around us could have originated as a vacuum fluctuation—essentially from nothing at all—because the large positive energy of the masses in the universe can be counterbalanced by a corresponding amount of negative energy in the form of the gravitational field. 'In my model' wrote Tyron, 'I assume that our Universe did indeed appear from nowhere about 10,000 million years ago. Contrary to popular belief, such an event need not have violated any of the conventional laws of physics'.

If Tyron is correct (and as a simple statistician I have to leave this for others to argue) it means that the Universe and everything in it was produced from a quantum fluctuation. The quantum fluctuation is a well-known phenomena—it triggers the emission of a quantum of light by an atom and the radioactive decay of a nucleus. It is an unpredictable, random event with no cause and if it can explain the Big Bang, chance really does appear to rule. In this case, the 'Grand Design' so dear to the hearts of theologians throughout the ages, is apparently reduced to a glorified cosmic hiccup. A random tear in the preexisting grand symmetry of nothingness.

"WE NOW KNOW ALL THE EXTRAORDINARY CHANGES THE UNIVERSE WENT THROUGH IN ITS FIRST SECOND. AFTER THAT, UNFORTUNATELY, IT TURNS OUT TO BE VERY MONOTONOUS."

But even if Tyron is wrong and, say, Greek mythology is right (and I suspect that many readers might be more convinced by the latter than the former), we cannot escape the conclusion that chance has played and continues to play a major role in the construction and evolution of the Universe in which we live— and presumably in the many other parallel Universes that some physicists consider to be a necessary consequence of quantum theory. And in our everyday lives chance regularly intervenes, often mercilessly, but on occasions with a more kind-hearted face. It seems that Anatole France may have hit the nail right on the head when he cheekily remarked that 'Chance is the pseudonym of God when he does not want to sign'.

I sense that I am rapidly getting out of my depth here—talk of God and parallel Universes should be left to theologians and philosophers. Statisticians are a humbler form of life, spear carriers rather than crown wearers, and we shouldn't forget it. So I shall now refrain from any more flamboyant speculation about life, the Universe and everything and instead amble down to my local bookie's to take my chances on this afternoon's card at Sandown Park. As my old uncle used to say, 'Be lucky'.

Epilogue

On reflection I thought someone a little more grand than my old uncle should have the last word. And so...

> The fabric of the world is woven of necessity and chance. Man's reason takes up its position between them and knows how to control them, treating necessity as the basis of its existence, contriving to steer and direct chance to its own ends.
>
> J.W. Goethe, *Wilhelm Meister's Apprenticeship*

Sources

Many books are mentioned in the text; details of these are given below along with a number of other texts which, although not directly referred to, acted as a source of some of the material in various chapters.

Games, Gods and Gambling, F.N. David, 1960, Griffin, London.

Chance in Nature, Edited by P.A.P. Moran, 1979, Australian Academy of Science.

Follies and Fallacies in Medicine, P. Skrabanek and J. McCormick, 1989, Tarragon press, London.

Life: An Unauthorised Biography, R. Fortey, 1997, Harper Collins, London.

Theories of Everything, J.D. Barrow, 1991, Claredon Press, Oxford.

The Inflationary Universe, A.H. Guth, 1997, Jonathan Cape, London.

The Blind Watchmaker, R. Dawkins, 1986, Longman Scientific and Technical, London.

The Man who Loved Only Numbers, P. Hoffman, 1998, Fourth Estate, London.

A Mathematician Reads the Newspapers, J. Paulos, 1993, Penguin, London.

Innumeracy, J. Paulos, 1994, Penguin, London.

A Mathematician Plays the Market. J. Paulos, 2003, Penguin, London.

The God Delusion, R. Dawkins, 2006, Bantam Press, London.

Mathematics of Chance, J. Andel, 2001, John Wiley, New York.

The Quark and the Jaguar, M. Gell-Mann, 1994, Abacus, London.

Taking Chances, J. Haigh, 2003, Oxford University Press.

The Making of the Fittest, S.B. Carroll, 2008, Quercus, London.

Index

Printed in the United States of America